Ronald A. Fritzsche

D0557934

Writing for Fishery Journals

Publication of *Writing for Fishery Journals* was supported with equal generosity by the

Fisheries Management Foundation
Seattle, Washington

and the

International Association of Fish and Wildlife Agencies
Washington, D.C.
William A. Molini, President

This book is the outgrowth of a symposium developed and sponsored by the American Institute of Fishery Research Biologists. The symposium was held September 13, 1988, during the Annual Meeting of the American Fisheries Society in Toronto, Ontario, Canada.

John Hunter, *Moderator*

Writing for Fishery Journals

Edited by
John Hunter

American Fisheries Society
Bethesda, Maryland
1990

Copyright by the American Fisheries Society, 1990
Library of Congress Catalog Card Number: 90-81939
ISBN 0-913235-65-2

Address orders to

American Fisheries Society
5410 Grosvenor Lane, Suite 110
Bethesda, Maryland 20814, USA

Contents

List of Fish Species

The colloquial names of most United States and Canadian fish species have been standardized in *A List of Common and Scientific Names of Fishes from the United States and Canada,* Fourth Edition, 1980, American Fisheries Society Special Publication 12. Throughout this book, species covered by the *List* are cited by their common names only. Their respective scientific names follow.

Alewife *Alosa pseudoharengus*
Bluegill *Lepomis macrochirus*
Brook trout *Salvelinus fontinalis*
Brown trout *Salmo trutta*
Chinook salmon *Oncorhynchus tshawytscha*
Coho salmon *Oncorhynchus kisutch*
Dolly Varden *Salvelinus malma*
Fantail darter *Etheostoma flabellare*
Fountain darter *Etheostoma fonticola*
Gulf menhaden *Brevoortia patronus*
Lake trout *Salvelinus namaycush*
Largemouth bass *Micropterus salmoides*
Lollipop darter *Etheostoma neopterum*
Pacific ocean perch *Sebastes alutus*
Pink salmon *Oncorhynchus gorbuscha*
Plains killifish *Fundulus zebrinus*
Rainbow trout *Oncorhynchus mykiss*
Sculpins: Family Cottidae
Sea lamprey *Petromyzon marinus*
Steelhead *Oncorhynchus mykiss*
Striped bass *Morone saxatilis*
Teardrop darter *Etheostoma barbouri*
Walleye *Stizostedion vitreum*
Walleye pollock *Theragra chalcogramma*

Foreword

The irritability of editors is legendary among authors. The capacity of authors to irritate is renown among editors. Much has been written about the pain that editors inflict on authors. This book is a report from the editors' ward.

To be sure, editors grate on authors. We criticize, make tart comments, ask ignorant questions, misconstrue content and intent, and make severe demands. We also invest a lot of effort in getting good work published as soon as possible, and we become annoyed and frustrated—and overly candid—when authors make this process difficult. Few authors intend to cause problems, and a few work hard to avoid them. Most are simply unaware that what they may have done (or not done) has made an editor's life unnecessarily difficult, and they are unprepared for the intensity of the editorial response they receive.

At the level of publishing that counts—dissemination of information—authors and editors are allies, not adversaries. Many mutual irritations can be reduced if each party understands the constraints on the other and takes steps to accommodate them. With this in mind, the American Institute of Fishery Research Biologists convened, in September 1988, a panel of current and past editors of fishery journals to relate their experiences, problems, and solutions to an audience of appreciative authors. Organized by John Hunter with early help from Eric Prince, and held in conjunction with the American Fisheries Society's annual meeting, this symposium was titled: "Writing for Fishery Journals: Pet Peeves of Editors and How to Avoid Them." The editors, as it happened, were anything but peevish; their emphasis was on improving the professional stature of authors, not on editorial politics. Their topics embraced strategies and procedures of publishing, statistical analysis of data, graphical and tabular displays of information, and word use and style. Most of the presentations, revised for publication, follow in this book.

Because these editors have (mostly) sublimated their annoyances, I should try to explain why manuscripts often become more acidic as they pass through journal offices. First, editors are as busy as everyone else. Many are volunteers whose wages are paid by other masters. Even we professional editors have more projects than we can handle comfortably. Any time-consuming problem that would not have arisen if an author had taken more care in manuscript preparation is a problem resented. Second, boredom—not with content but with recurrent nuisances—is an editor's constant companion. An author may have dangled her first participle or garbled his first discussion, but the editor has had to fix these problems dozens or hundreds of times already. Everyone is entitled to a few slipups per paper, but editorial patience wanes rapidly as mistakes accumulate. Third, most editors have more editorial responsibilities than usually is appreciated, and an author can get crosswise with all of them. These responsibilities sort broadly into three categories: quality control, cost control, and communication.

• Quality control of technical information is the best known and (usually) respected function of journal editors; its most obvious manifestation is peer review. Altruism motivates this function, but so does knowledge that an editor's professional reputation is as closely tied to journal quality as an author's. An editor cannot tolerate substandard technical work gladly. If a paper is so poorly organized and written that it takes a major effort to discern its technical merit, editorial barbs may be especially sharp.

• Journals are expensive to publish and must operate within strict budgets. Editors cannot allow long-windedness, excessive data and details, or badly designed figures and tables to consume space and run up costs. If they do, fewer papers can be published, author and then reader loyalty will erode, and the journal may fail. Cost also underlies many of a journal's (and editor's) idiosyncracies of style and procedure. For each publisher, each editor, each typesetter and printer, some ways of doing things are more efficient and cost-effective than others. Authors can never wholly decipher or outguess an editorial system (which continually evolves anyway), but they can impose major costs on it by ignoring a journal's conventions altogether.

• Good editors root for their authors but cater to their readers. There is no journal without an audience; the audience will be meager if the journal is intelligible only to a few specialists. Editorial edicts about grammar, syntax, clarity of expression, jargon, definitions, table and figure design, and more all serve the journal's (and the profession's) communication function. If some of the edicts seem strange or unreasonable, they probably reflect the editor's acquired knowledge of standards and conventions throughout the profession and in related ones—a perspective that specialized authors may lack. However important the edicts may be, they become repetitious and tedious to issue, and may not reach an author with grace.

With this very personal overview of editorial context and problems, I defer to my colleagues for their practical advice to authors. I am grateful to John Hunter and the AIFRB for providing the message, and pleased to have played a part with the medium.

Robert L. Kendall
Managing Editor
American Fisheries Society

Usage and Style in Fishery Manuscripts

PAUL H. ESCHMEYER

U.S. Fish and Wildlife Service
Office of Information Transfer
1025 Pennock Place
Fort Collins, Colorado 80524, USA

Abstract.—Fishery manuscripts of the U.S. Fish and Wildlife Service have been routinely reviewed in-house, before their submission for publication, to help ensure the effectiveness of presentation and the clarity and appropriateness of language. Details of English usage and style associated with common writing problems, and that are briefly discussed herein, include confusing word pairs (*affect, effect*; *principal, principle*), awkward constructions (backward sentences, stacked modifiers), excess baggage (*activity, operation, located*), interruptions of forward motion (*former, respectively*), polysyllabism (*approximately, subsequently*), subjects of disagreement among language authorities (*likely, parameter*), and others. Some of the principal problems have changed little over the years; many of the topics treated also appeared in informal memoranda from Service editors to fishery scientists in the then U.S. Bureau of Commercial Fisheries in 1965–1969. The principal admonition throughout has been that fishery writers who are communicating their research results to the scientific community and committing them to the permanent record have a strong obligation to write succinctly in clear, simple, standard English and to avoid usages that impede readability and comprehension.

Fishery scientists are not alone in having writing problems. Good writing is a formidable undertaking for nearly everyone. The stress is reduced in well-trained and experienced journalists and others who write continually, but is often severe in scientists who write for publication only occasionally, when reporting the results of their research. For them, writing is often one of the most difficult tasks in their professional careers, far removed from the implication in the offhand comment heard in laboratory hallways: "Yes, I've finished the research. All I have to do is write it up."

And "write it up" they do. Many who have had little practice and often inadequate training (Hile 1971) simply start writing, improvising ineffectively as they write. Inevitably they are beset by the pitfalls of faulty grammar, usage, and style.

Some of the problems they encounter are briefly discussed here and illustrated by examples taken mainly from unedited manuscripts prepared by fishery scientists of the U.S. Fish and Wildlife Service (here averred to be written, on the average, as well as or better than manuscripts prepared by other groups of fishery scientists). Service editors have routinely reviewed these manuscripts to evaluate the clarity and appropriateness of language and style and to offer suggestions for improvement, before the papers are revised by the authors and sent to journals or other publication outlets.

These detailed in-house reviews began in 1965, when an editorial system was established in the then Division of Biological Research of the then U.S. Bureau of

Commercial Fisheries by Ralph Hile,[1] with my assistance. As part of our communication with fishery writers in the Division, we periodically issued informal one-page *Editorial Memos* that dealt with writing faults in manuscripts we had recently reviewed.

The reappearance here of some subjects treated in the memoranda of 1965–1969 suggests that many problems in fishery writing have changed little over the years. This persistence of problems is not unique to the fishery field but has extended throughout written and spoken English over much longer periods—as demonstrated by the continued use of certain long-published works on writing. Two prominent examples are "the *little* book" of William Strunk, Jr., *The Elements of Style*, first copyrighted in 1918 and still an excellent reference in the 1980s, and H. W. Fowler's *A Dictionary of Modern English Usage*, first published in 1926 and (in a revised but closely similar 1965 edition) still one of the most useful and up-to-date references available. A dozen or so other major works that concern hundreds or thousands of details of English usage have been published in more recent years, particularly 1957–1987.

The many subjects treated in these authoritative works cover nearly all aspects of written and spoken English. However, the often lengthy and detailed accounts are sometimes difficult to use efficiently, comparatively few of the topics are of frequent concern to the writing scientist, and some terms used in fishery writing are not included.

The treatments offered here are intended to provide easy access to information on certain details of usage and style that have caused frequent or conspicuous problems in fishery writing. A few reflect only personal observations, but most are supported by citations of major references on writing that should enable fishery scientists to easily increase the scope of their search for information on the subjects treated.

The accounts that follow in alphabetical order concern the use of words (here in italics) and matters of style, readability, and mechanics: abbreviations; accuracy and consistency; acknowledgments; *affect, effect, impact; and/or*; backward sentences; beginners' bloopers; *between, from*; *case*; *comprise*; conformity with publisher's style; *data* et al.; Discussion; egocentric action words; excess baggage; *fisheries, fishery*; fish names; *former, latter*; *like, likely*; *occur*; overspecification; *parameter*; polysyllabism; *principal, principle*; punctuation; *quite*; *relation, relationship*; *reside*; *respectively*; spelling; split infinitives and compound verbs; stacked modifiers; straining to be scholarly; *that, which*; *use, using, utilize, utilization*; weasel words; *while*; and *with*.

---

[1]Dr. Ralph Hile, a premier fishery scientist of this century, was a Fellow of the American Institute of Fishery Research Biologists and an Honorary Member of the American Fisheries Society. He was Editor of the *Transactions of the American Fisheries Society* in 1945–1948, President of the Society in 1952, and recipient of the Society's Award of Excellence in 1972. During a distinguished career of nearly 40 years in the U.S. Government, he produced fundamental and pioneering works on such diverse subjects as morphological variation in coregonids; age, growth, abundance, and distribution of freshwater fishes; and methods of statistical analysis and biological interpretation of catch-effort records of the commercial fisheries of the Great Lakes. He was Biological Editor of the Division of Biological Research from 1965 until his retirement in 1969.

Style and Word Use

Abbreviations

Writing is often eased and readability improved by the judicious use of abbreviations composed of the initials of long words or names of organizations—as in the replacement of *dichloro-diphenyl-trichloro-ethane* with *DDT* or *Food and Agriculture Organization* with *FAO*. As might be expected, these potentially useful and convenient aids to communication are subject to misuse and overuse. The most devastating misuse is the insertion of an unfamiliar set of initials that is defined nowhere in the manuscript. An exasperated reader is left guessing the intended meaning—often unsuccessfully. At the opposite extreme is the careful introduction and definition of an initialism, alerting the reader to recognize and perhaps memorize it, that never reappears. Sometimes many more are introduced within a short span of text—typically in the Methods section—than an overwhelmed reader can possibly absorb. Careful judgment is required to control overuse; not only the length and cumbersomeness of the name or term being abbreviated must be considered, but also the frequency of its use in the paper. A considerate writer restricts the use of initialisms to those that contribute significantly to readability, introduces each one at first use in the text, and (in a long paper) provides an easily accessible list or table defining the potentially troublesome nonstandard abbreviations used in that work.

Accuracy and Consistency

Too often too little attention is given to purely mechanical details in the final stages of preparation of fishery manuscripts. Careless errors undermine the confidence of reviewers, who are understandably led to wonder whether the errors might also reflect the quality of the research being reported.

The author has the primary responsibility for undertaking the meticulous point-by-point scrutiny required to ensure accuracy and consistency in a manuscript. Although even the most painstaking work cannot be expected to produce a perfect paper, patience and perseverance can spare an author the embarrassment of entering errors into the permanent record. Word processors and copying machines have reduced the chances of new errors creeping into successive drafts of a manuscript, but the factor of human fallibility remains. Certain routine procedures listed here, applicable especially to the final draft, help reduce the incidence of errors.

• Read through the text slowly, and check each statement against the relevant figure, table, or other source. Ensure that there are no discrepancies between the text and the tables, figures, or Abstract.

• Check each direct quotation against the original to ensure accuracy.

• Check the spelling of scientific and geographic names and of unusual terms.

• Check every citation in the text against the corresponding entry in the References to ensure complete agreement.

• In the References, ensure that the style details of the citations listed agree with those in recent issues of the publication to which the manuscript is to be submitted.

• Check each paper listed in the References against the original publication to ensure the correctness of all details of the citation: authors (restricted to the initials and surname in nearly all fishery works); year; title, unless omitted in the style of the intended publication; name and, when applicable, the volume or number of the publication; and pagination. Careless errors that are painfully common in the titles of references are the omission or addition of punctuation and the omission or misspelling of scientific names or the authors of scientific names (the misspelling "Mitch*e*ll" for "Mitch*i*ll" is by far the most frequent example).

• Check galley or page proofs with great care; few are free of error. Do not revise your paper in proof, and especially avoid making unnecessary changes in page proofs.

• In the printed paper, be prepared philosophically to see a glaring error on the first page you examine. It was ever thus.

Acknowledgments

One point of view is that Acknowledgments are personal expressions of appreciation from authors to coworkers and others and should be left written in the authors' style and wording. But there is room for disagreement. Too often one sees examples of scholarly, highly formal writing followed by an Acknowledgment section in which the author shows clear signs of intellectual and emotional collapse. Discordant notes are struck by fond expressions of profound thanks to Shorty Miller and deepest obligations to Slim Smith for their excellent help in the laboratory, and undying gratitude to Molly Coddler for the superb typing of the manuscript. And there must be a more appropriate place for expressing such sentiments as, "Finally, I must thank my family for putting up with the tremendous amount of my time they had to get along without over the long grind."

Often the most effective acknowledgments are simply stated, as in "We thank A. B. Miller and C. D. Smith for assisting us in the laboratory and E. F. Brown and G. H. White for reviewing the manuscript." In the frequent "We wish to thank" opening, the wish is easily fulfilled if the extra words are merely omitted. Initials alone work well, and some journals insist on them; overly fond or descriptive nicknames do not. Alphabetical sequence within the groups of persons acknowledged seems more evenhanded than a random arrangement of names.

affect, effect, impact

Among similar-looking pairs of words that have completely different meanings and that have been confused for centuries, *affect* and *effect* may be the most troublesome to fishery writers; however, the words can be held within their respective spheres of meaning. It is safe to say, for example, that if the intended word is a noun, *effect* is the word needed; the noun *affect* is confined to technical uses in psychology and related fields and is unlikely to appear in a fishery manuscript. As a verb, by far the most often used member of the pair is *affect* (influence, have a bearing on, as in "Water quality affects the well-being of fish.") The word is almost never used in the sense of pretense or imitation in fishery writing. The verb *effect* means accomplish, cause, bring about, as in "The second application of rotenone effected a complete kill." Although there may be no foolproof way of teaching the distinction between the two verbs, Bernstein (1977) tried: "Think of the first letter of each word—the *a* in *affect* and the *e* in *effect*. Then think of the *a* as standing for *alter* (which is not a synonym for *affect* but is close to the meaning) and the *e* as standing for *execute* (which ditto). Now recite the previous sentence 100 times. . . .''

In fishery manuscripts, by far the most common miscues are the use of *affect* as a noun where *effect* is intended, and of *effect* as a verb where *affect* is intended.

But even while *affect* and *effect* wrongfully usurp each other's rightful places, the relative newcomer *impact* too often encroaches on both words. Although *impact* long was a forceful word that kept company with *collision*, *concussion*, and *impingement*, misuse and overuse—no doubt stimulated by the advent of the Environmental Impact Statement and wide acceptance by government agencies—have reduced its strength and in some usages pushed it strongly toward synonymy with *affect* and *effect*. Too often one sees tedious statements about a fishery impacting a fish population or having an impact on it, or a fish population being negatively impacted by a fishery (or even that ''improving the work space can impact people and productivity''). Such usages have been vigorously denounced by authorities. Barzun (1975) listed *impact* among ''good words diverted from their plain sense . . . '' and advised that ''If you have to report the *impact* of a meteorite in your backyard, use the word without hesitation, but forgo it in describing the effect of a new detergent on the lives of young mothers.'' Other authorities have regarded the word in some of these uses as gobbledygook (Copperud 1964), faddish, used by those ''who are always reaching for the flame thrower to light a cigarette'' (Bernstein 1965), voguish (Follett 1966), or barbaric (Kilpatrick 1984). Nevertheless, such uses are now recognized as ''entirely standard'' by the *Random House Dictionary of the English Language*.

But standard or not, it seems certain that too many knowledgeable readers have seen too many inept uses of *impact*—''The ear flinches and the hair curls'' (Kilpatrick 1984)—as either verb or noun, where *affect* or *effect* would serve better. The careful fishery writer simply avoids using *impact* in the widely criticized senses.

and/or

Few words or word combinations commonly seen in fishery manuscripts have exceeded *and/or* in eliciting the disapproval of authorities on usage. Strunk and White (1959) wrote that *and/or* is ''Useful only to those who need to write diagrammatically or enjoy writing in riddles''; others have called it an ''ugly device'' (Fowler 1965), ''a visual and mental monstrosity'' (Bernstein 1965), and an ''ungraceful expression'' (Follett 1966). It has been relegated primarily to legal and business writing. But even in the legal profession it has not always been warmly received, having been described by a Justice of a State Supreme Court as ''that befuddling, nameless thing, that Janus-faced verbal monstrosity, neither word nor phrase, the child of a brain of some one too lazy or too dull to express his precise meaning, or too dull to know what he did mean . . .'' (Fowler 1935). Clearly, *and/or* should not properly be part of careful fishery writing, for ''It destroys the flow and goodness of a sentence'' (Strunk and White 1959).

Backward Sentences

A substantial part of editorial work consists of transposing phrases to bring the subject closer to the start of a sentence. Many authors seem reluctant to disclose the topic early in the sentence and a few avoid direct statements with astonishing consistency. The introductory phrase is not invariably bad, for it is commonly needed to give the correct sequence of ideas or to balance a sentence. Excessive use or excessive length of the windup opener, however, makes a manuscript dull and hard to read. The kind of rhetorical throat-clearing that precedes the significant part of the sentence varies, but must surely include "During . . ." and "There was . . .," taken separately, or together—as in "During the months of July and August there was an increase in mortality" (to be compared with "Mortality increased in July and August"). Other disclosures of the point being made are more elaborately delayed: "Taking into consideration the length of time the collecting gear was in place (6 months, from 1 May to 30 October 1985), as well as available evidence regarding the time when the animals spawn, it was possible to predict that. . . ."

In one of the most onerous and frequently seen forms of the backward sentence, a reader must plow through a long series of references without having the slightest notion of what the sentence is about. An example is the opening sentence of a published fishery paper (not verbatim here, but not exaggerated): "Recent research (Aardvark et al. 1985, 1986a, 1986b, 1986c; Aardwolf 1986; Aarhus et al. 1986; . . .[seven additional name-and-year citations given]) suggests several possible methods. . . ."

Equally fatiguing or exasperating to a struggling reader is a long series of scientific names preceding the verb: "Six species of planktonic diatoms (*Asterionella formosa*, *Cyclotella comata*, *Fragilaria crotonensis*, . . .) and one benthic diatom (*Achnanthes minutissima*) were common in the river in 1980–1983." In a more pleasing style the subject is brought to the front of the sentence: "Forms common in the river in 1980–1983 were six planktonic diatoms . . . and one benthic diatom. . . ."

Beginners' Bloopers

Beginning or careless writers—or perhaps those hurrying pell-mell to meet a tight deadline—sometimes commit embarrassing miscues or oversights that might be interpreted by an editor, reviewer, or reader as evidence of rank amateurism. Listed here, in no particular order, are some that appear often enough in fishery manuscripts to warrant mention.

- Including literature citations in the Abstract.
- Introducing unused or unneeded initialisms in the Abstract.
- Citing in the text a personal communication from a coauthor of the paper, or mentioning the help of a coauthor of the paper in the Acknowledgments.
- Using the first person singular *I* in manuscripts with more than one author, or *we* in manuscripts with a single author.
- Listing in the References a paper, with the name of the journal or other hoped-for publication outlet, that has merely been submitted for review but has not yet been accepted (or rejected).
- Beginning sentences with abbreviations, as in "*D. magna* was sensitive to selenium."
- Italicizing a family name (*Centrarchidae*), or the abbreviation for *species* (*Perca sp.* or *spp.*), or the authority's name (*Perca flavescens Mitchill*); failing to capitalize a family name (centrarchidae); or capitalizing a specific epithet (*Perca Flavescens*) or the vernacular form of a family name (Centrarchid).

• Mistreating the most common and useful Latin abbreviations: using *et* as an abbreviation in literature citations, as in "Jones et. al. (1996)"; carelessly confusing *e.g.* (for example) with *i.e.* (that is); writing the clearly redundant "and etc."; and using *e.g.* and *etc.* together, as in "e.g., seines, trawls, gill nets, etc." where the abbreviations doubly advise the reader that the list is incomplete (here, "such as seines, trawls, . . . etc." is equally disconcerting).

• Assigning age 0+ to young-of-the-year fish without explanation, or routinely affixing clumsy plus signs to fish ages where they contribute nothing to clarity or understanding.

• Labeling the *Y*-axes of graphs upside down (the lettering should read from bottom to top—never from top to bottom).

• Offering incongruous mathematical relations, as in "Concentrations lethal to daphnids were 1,000 times lower than those lethal to midges."

• Using mathematical signs or symbols carelessly—e.g. (the primary offense), $P \geq 0.05$ when $P \leq 0.05$ is intended, or vice versa.

• Submitting a single-spaced manuscript to a reviewer or editor.

• And potentially the most exasperating of all, failing to number the pages of a manuscript. The reviewer in whose care the pages of a long, unpaginated manuscript become accidentally scrambled will remember the author forever.

between, from

A small but significant fault commonly seen in fishery writing is the use of an en dash or hyphen to separate numerical intervals introduced by *between* or *from*. In conventional English (not including the utterances of Madison Avenue or accounts in the sports section of the daily newspaper) *between* requires a following *and* (not *to* or a hyphen), as in "the fish weighed between 6 and 8 kg (not "between 6 to 8 kg" or "between 6–8 g"); and *from* requires a following *to*, as in "from 1985 to 1988" (not "from 1985–1988," a construction properly regarded as illiterate by the literati).

case

Although the word has legitimate usage in references to patients in medicine and to arguments in law, *case* is rarely useful in the open text of fishery writing. One needs only to compare "Examination of 107 females revealed 10 cases of egg retention" with "Of 107 females, 10 had retained their eggs." The useless use of *case* has been vigorously criticized by authorities on writing: "Despite its score of legitimate uses, *case* is continually being used when it has practically no meaning at all . . ." (Evans and Evans 1957); "There is perhaps no single word so freely resorted to as a trouble-saver, and consequently responsible for so much flabby writing" (Fowler 1965); "*Case* often forms the nucleus of a foggy phrase, which can either be dropped or replaced by something more specific" (Bernstein 1965); and "One should never use *in case of* to announce what a sentence is about . . ."— rather, total abstinence is encouraged until the lazy reliance on the word is broken (Follett 1966).

comprise

In conventional English, the whole comprises the parts ("The family comprises 20 species") and the parts compose, constitute, make up the whole ("The family is composed of 20 species''). By this criterion, *is comprised of*, as in ''The family is comprised of 20 species''—the form seen most often, by a wide margin, in fishery manuscripts—is invariably wrong. Nevertheless, the attractive, perhaps learned-sounding *comprise* is irresistible in scientific writing, compared with the mundane *compose* (which seemingly is never used where *comprise* would be preferred). As an inevitable consequence of the long-time use of *comprise* for *compose* (since the late 18th century, according to *Webster's Ninth New Collegiate Dictionary*), the two words are moving toward synonymy or have already arrived. In scientific writing, however, the prudent author continues to follow the grammatically unassailable approach of preserving the distinction between the two words.

Conformity with Publisher's Style

Few writing faults are as highly visible or as prevalent as the failure of authors to tailor their manuscripts to match the style of the intended outlet. The preferences of different publishers differ in many ways—some major, some minor. Responsible writers read and carefully follow the ''Guide for Authors,'' ''Information for Contributors,'' or equivalent instructions that usually appear in the first yearly issues of journals (or more often), or are readily available from editors or publishers; they consult the *CBE Style Manual* (CBE Style Manual Committee 1983) for matters not specifically treated in such instructions; and they infer other details from usages in recent issues of the publication concerned. Such writers set themselves above the crowd. They avoid the inevitable embarrassment and loss of time that results when manuscripts are returned out of hand, for nonconformance with style, even without review; they reduce the work required later to prepare a revision; they expedite the appraisal of manuscripts by reviewers and editors; and they gain the very considerable appreciation of reviewers, editors, copy markers, and typesetters.

data et al.

Opinions about the use of *data* as a singular or plural were about equally divided among critics surveyed by Copperud (1980) and among usage panelists of the *American Heritage Dictionary* and Morris and Morris (1985). However, *Webster's Ninth New Collegiate Dictionary* noted that ''Our evidence shows plural use to be considerably more common than singular use.'' And *data* remains strictly plural in fisheries and throughout the natural and physical sciences. Thus ''the data is'' is regarded by most fishery readers as a blatant error and should be meticulously avoided. Even when the word is correctly used with a plural verb (''The data are . . .''), a point often overlooked is the inappropriateness of singular modifiers (*little*, *much*, *this*) and the requirement for plural modifiers like those used for other countable items—*few*, *many*, or *these*—as in ''Few of these many data are legible.''

The plural forms of at least four other words from Greek or Latin—*criteria*, *media*, *phenomena*, *strata*—are now commonly treated as singular (for example, ''the media is'' is surely heard and seen far more often than ''the media are''), but they remain solidly plural in scientific writing. An exception is *agenda*, which has fully made the transition to the singular in standard English. As Bernstein (1965) noted, ''*Agenda* has departed from its original meaning of things to be done, and now means a program of things to be done. Its singular force is so strongly felt that the word has developed its own plural—*agendas*.''

Discussion

Long use and convenience have imposed the indicative subject headings "Results" and "Discussion" firmly on the organization of fishery research papers. Their use, together with "Introduction" and "Methods," is specifically called for by the "Instructions to authors" published by many journals and in some books on scientific writing. But that admonishment is too stringent. Although these headings provide convenient compartments for the arrangement of information in many reports, some others—especially multiple-theme papers—are better handled by dividing them into sections with informative headings. Experienced scientists allow the materials at hand to guide the organization of a paper, and follow the stereotyped Results–Discussion style only if it best meets their needs.

Many fishery writers who use these indicative headings introduce their work satisfactorily, describe their Methods adequately, and arrange their Results in logical order, but then become overwhelmed by the Discussion. To some, this all-inclusive heading is an invitation to wax verbose, to expound at length and in detail, and perhaps to pontificate a bit. A frequent fault is the verbatim or nearly verbatim repetition of statements from the Results section, which the authors believe to be necessary to formally bring each topic up for discussion. A special example is provided by the converted thesis, in which the writer successively examines various possibilities in detail and then in equal detail shows that they could not possibly apply to the work at hand. The result is a distinctive roller-coaster pattern produced by materials that perhaps once helped convince a graduate committee that the subject had been exhaustively examined, but that have no place in a professional paper.

No panacea is offered here. The Discussion is often the most difficult section to prepare and calls for careful thought and concise, coherent composition. The addition of informative subsection headings often helps both the writer and the reader concentrate on the specific aspect of the subject being discussed.

Egocentric Action Words

Scientific writing can be injured through excessive use of action words, whereby the writer directs and redirects the reader's attention to the investigator's activities. In the following examples, the italicized words take up space but seem to contribute little: "All of the 22 fish *encountered* were *identified as* . . .''; "The fish *collected* were *found to be* . . .''; "*Observations revealed that* . . .''; "*The results of this study showed that* . . .''; and so on. Not all action words are in fact superfluous; those that are can be easily identified by a writer who is on guard against the evil.

Excess Baggage

Fishery writers must constantly be on the watch for lurking freeloaders that seem eager to hang out with working words but merely occupy space and weaken sentences. Four of these ever-ready intruders that shed no light are *activity*, *effort*, *operation*, and *program*. The question to be asked is whether "We completed the research activity (effort, operation, program) in September" is in any sense an improvement over "We completed the research in September" or whether "The crew began the sampling operation (activity, effort, program) in June" outperforms "The crew began sampling in June." An affirmative answer is usually insupportable.

Other inert fillers similarly lie in wait for a chance to usurp space, detract from succinctness and readability, or commit other atrocities. The following sentences demonstrate the offense. Mere excision of the offending words—here enclosed in parentheses to facilitate the surgery—often improves the prose: "Adverse weather (conditions) kept the vessel in port"; "The temperature had risen to (the) 20°C (level)"; "The salinity (value) was 34‰"; "The plankton distribution (pattern) was patchy"; "The headwaters of the stream are (located) in Mercer County, Ohio"; "Sampling sites were established (at locations) 50 and 300 m upstream"; and "The power plant is (situated) 2 km from the dam."

fisheries, fishery

Should one write "fisheries management" or "fishery management"? The question has generated considerable heat for decades, and the subject still sometimes flares up unexpectedly.

Historically, *fishery* was the first of the two words recorded by the *Oxford English Dictionary* as being used attributively, in a quotation dating back to 1528: "The fyssherye house at Guisnes." The first use of *fisheries* quoted there was "Fisheries Exhib." in 1883—just one year before the American Fisheries Society adopted its present name. Thus *fishery* and *fisheries* have been used as modifiers in parallel constructions for more than a century.

Among more recent authorities, Evans and Evans (1957) wrote that "Almost any English noun can be used in the singular form as if it were an adjective" and "Traditionally, the singular form is required for the first element in a compound. . . ." They added, however, that "in current English, especially in the United States, a good many plural forms are appearing in this construction, as in *communicable diseases control, welfare services funds, correctional institutions specialist.*" The implication was that the plural element is acceptable because it "offends very few people when it is used in heavy, unfamiliar compounds made up of abstract ideas."

Morris and Morris (1985) offered a different perspective in reply to a correspondent who complained about the "ungrammatical, non-euphonious, and obnoxious" use of pluralized adjectives in terms such as "resources conservation," "fisheries management," and "lands acquisition." These authorities wrote in reply that "the examples cited were not adjectives used as plural modifiers" but actually plural nouns used attributively, and labeled the usage as a grievous example of the "mismating of plural nouns and singular nouns." Their analysis seemingly permitted the inference that one may write "fisheries resources" but not "fisheries resource" (which the authors would surely include among "mismatched plurals and singulars from the language of Washington bureaucrats"). But this position has not been supported by usage in scientific writing in the United States.

Everything considered, the answer to the question posed here must be that either *fishery* or *fisheries* is fully acceptable as a modifier in all usages—*fishery* because it is grammatically irreproachable and *fisheries* because it has become securely established by long use.

The plea remains that, in contradiction to the "mismatched plurals and singulars" of Morris and Morris (1985), either the singular or plural form be used consistently in a given manuscript, to avoid the sense of irregularity or indecision that seems to result from the use of a mixture of the two forms and that tends to impair readability. (Readability is not affected by exceptions that must be made for the formal names of agencies or organizations, such as the Michigan Institute for Fisheries Research or the Oregon Cooperative Fishery Research Unit, because the reason for the exceptions is immediately evident.)

Fish Names

Fishery workers in Canada and the United States are fortunate in having an authoritative, long-established list of common and scientific names of fishes, now in its fourth edition (Robins et al. 1980) and approaching its fifth (Robins et al., in press). This work has helped immeasurably to promote uniform usage and avoid confusion.

Although adherence to the list is seldom questioned, the use of partial names rather than full common names in fishery manuscripts is a recurring problem. Many prefer to write "coho," "pink," or "rainbow" rather than "coho salmon," "pink salmon," or "rainbow trout." Often the shortened names sound colloquial and inadequate, and the plurals bizarre—e.g., "22 pinks and 15 rainbows." And if exceptions are made for one group they can hardly be denied for another, and widespread exceptions to full common names must lead to chaos. If the omission of "salmon" or "trout" from common names is accepted, a researcher on darters presumably has equal right to record the collection of "teardrops," "fantails," "fountains," or "lollipops."

Careful writing can lessen the repetition of a common name that seems bulky. If a paper is concerned with a single species of trout or darter, the group name can sometimes stand alone, provided the first common name in a paragraph or subsection is complete and the identity of the species remains unmistakable. Or, if "22 coho salmon" causes distress, one can write (when the subject species is clearly established) "22 fish," "22 adults," or "22 females."

Sometimes not so easily resolved by fishery writers is the selection of plural names for the subjects of their work. The problem begins with *fish* itself, which has two accepted plural forms, *fish* and *fishes*. *Fishes* is the form generally preferred for all uses by systematists but is used by most other fishery workers primarily to signify diversity in kinds or species, as in "collections of fishes and mammals," "the fishes of Lake Erie," "coarse fishes," "freshwater fishes," and so on. For other uses in fishery manuscripts and most fishery journals, the irregular plural *fish* is preferred by a wide margin, as in "We caught 22 fish in the trap net," "Hundreds of fish were killed by the chemical," or "Each predator consumes about 600 fish per year." Morris and Morris (1985) accepted both plural forms, but endorsed a panelist's comment that "The normal plural, of course, is *fish*."

Evans and Evans (1957) wrote that *fishes* is the oldest form of the true plural ("as in *five barley loaves and two small fishes*" in the King James Bible) and regarded the use of *fish* as a plural to be a recent development, probably derived from the long use of the singular *fish* generically or collectively or as a mass noun. Although allowing that "*fish* may now be used as a true plural," they wrote that "the movement toward regular plurals is very strong in modern English and the life expectancy of a new irregular plural, such as *fish*, is not very good." However, as judged by its long use in fishery publications and elsewhere, its life expectancy as seen from the perspective of the 1980s seems very good indeed.

The plural forms of the common names of some species of fish are difficult to deal with satisfactorily. The complexity of the problem is suggested in the front matter of *Webster's Third New International Dictionary*: "Many names of fishes . . . have both a plural with a suffix [trouts] and a zero plural that is identical with the singular [trout]. Some have one or the other. Some present a choice according to meaning or according to a special interest of the user." For species with regular plurals (i.e., with suffixes), the choice is clear and straightforward; nevertheless, some fishery writers and occasional fishery journals continue to use zero plurals for common species that clearly, by long use in conventional English or by explicit dictionary definition, require a suffix (alewives, bluegills, lampreys, walleyes), or to indiscriminately mix the correct plural forms with zero plurals in a paper, paragraph, or sentence.

For species with potentially irregular plurals, the principal suggestion here is that a writer consult a dictionary, as one might to determine the preferred plural of any other word in question. If the dictionary offers a seemingly even choice of plurals, a usually sound approach is to find and follow the usage that prevails in current issues of the *Transactions of the American Fisheries Society*.

former, latter

Few words rank higher than *former* and *latter* on the list of candidates for extirpation from the vocabulary of scientific writing. Almost invariably, they force the reader to stop, back up, and search—sometimes in vain—for the often obscure referent. *Former* is the more loathsome of the two, for it sends the reader back over greater expanses of verbiage to puzzle out the intended meaning. *Latter* almost equally often sends the reader backward, though usually on less lengthy searches over less distant reaches.

But *latter* has other encumbrances that make its use inadvisable. Usage authorities disagree about whether it may be used to refer only to the second of two things—seemingly the grammatically sound usage to which it should be restricted (if used at all) in scientific writing—or to the last-named of three or more or to the last three ("the latter three") or more of a longer list.

Careful writers simply construct sentences that avoid the troublesome words. They prefer to repeat the original words or phrases, if necessary, to ensure clarity and unbroken continuity, rather than resort to the use of either member of the pair.

like, likely

Scientists whose principal reason for writing is to report information—their research results—directly and unmistakably, have a special responsibility to write in the clearest, simplest standard English they can muster. Such writing implies an obligation to avoid questionable uses of words or constructions that may divert the attention of readers and thus impair readability. Both *like* and *likely* are in this category in certain usages.

Some fishery writers seem hesitant about writing *like* where it properly fits (as a preposition), as in "The largemouth bass, like most predaceous fishes, has a large mouth," and substitute the noninterchangeable *as*, "The largemouth bass, as most predaceous fishes. . . ." The substitution, which "sounds as hell" (Bernstein 1965), seems to be a holdover from the widely publicized criticism of an advertising slogan to the effect that "Surchil tastes good like a chocolate should" (or something like that). This use of *like* as a conjunction is criticized by some authorities and accepted by others. The consensus is that, though the usage is defensible, "the user should be prepared for criticism" (Copperud 1980). The prudent fishery writer simply avoids the need for such preparation.

The adverbial form *likely*, standing alone as an alternative for *probably*, is seen rather often in fishery manuscripts, as in "Sediments that fell on the reef were likely resuspended by the currents," or "The fish will likely die." This use of *likely*, unless it is preceded by an intensive such as *very*, *more*, *most*—as in "The fish will most likely die"—has been variously evaluated as having "an old fashioned tone" (Evans and Evans 1957), being unacceptable (Bernstein 1965; Morris and Morris 1985), and being "incurably awkward" (Follett 1966). Although Copperud (1980) wrote that "The consensus is that it [the use without an intensive] is correct," he noted that it sounds objectionable to people unaccustomed to it. The careful fishery writer, mindful of the need to keep the reader's attention on the subject and not diverted by the words, simply replaces *likely* with *probably*, against the use of which (to my knowledge) no disparaging word has ever been heard.

occur

Many fishery manuscripts are damaged by the overuse of weak words that contribute no real substance to sentences but only lengthen them and perhaps help hold them together grammatically. Unquestionably, *occur* is among the primary offenders, as illustrated in the following sentences, stronger versions of which, without *occur*, follow each in parentheses. "An increase in mortality occurred" ("Mortality increased"); "Captures of fish of similar size occurred in both traps" ("Fish of similar size were caught in the two traps"); and "Egg deposition occurs in October" ("Eggs are laid in October").

A double use of the word in a sentence more than doubles the damage: "A detrimental downstream movement of fry occurred during freshets if they occurred soon after fry emergence." The meaning is obscure. A possibility is that "Fry that had recently emerged were killed by early freshets that swept them downstream prematurely." However, their fate may have been much less tragic. Only the author knows.

Careful writers reexamine each first-draft sentence in which they have used *occur* to ensure that its removal would not be an improvement. In some contexts the use of the word is beyond objection: "The species occurs only at great depths off the Pacific Coast" and "Arsenic occurs in air, soil, water, and all living tissues."

Overspecification

Examples of the "tautological phrase of specification" of McCartney (1953) seen in fishery manuscripts include "few in number," "green in color," "cylindrical in shape,'' ''small in size,'' and so on. It is smoother to write, ''The fish was 60 cm long and weighed 4 kg'' than ''The fish was 60 cm in length and 4 kg in weight.''

In a different aspect of overspecification, the frequent repetition of details in the text often becomes unnecessarily tedious or painfully distracting. In a manuscript in which the authors noted in the Methods section that ''Total length (TL) of each fish was measured to the nearest millimeter,'' and in which no other type of length (e.g., fork, standard) was mentioned, *TL* was repeated after each of 22 lengths given in two pages of text.

Words that a writer has newly ''discovered'' (one suspects) or that sound singularly scientific, sometimes receive a vigorous workout; examples are 17 uses of *macrophyte* in three pages of typescript and 4 uses of *a priori* in nine lines. Too often the designated statistical limits of significance (typically $P \leq 0.05$ and sometimes also the seemingly superfluous $P > 0.05$) are duly repeated after many or most statements in the Results section. Considerate writers facilitate readability by stating the accepted level of significance in the Methods section, accompanied by the qualifying ''unless otherwise indicated'' where appropriate, and omitting it thereafter in the text.

Also disconcerting is the surprisingly frequent use of both *percent* and its symbol in a sentence that can accommodate only one or the other, as in ''The mean percent hatch was 32%,'' or ''The average percent recoveries were as follows: Ca, 108%; Mg, 59%;. . . .''

parameter

Among other authoritative sources, the *American Heritage Dictionary of the English Language* defined *parameter* as ''A variable or an arbitrary constant appearing in a mathematical expression, each value of which restricts or determines the specific form of the expression.'' Many fishery writers and others, however, use *parameter* as a collective term for characteristics of water (concentrations of ions and dissolved gases, temperature, transparency, and so on) or of mensural variables such as length and area—indeed for almost any characteristic or feature of almost anything. This all-inclusive usage is now accepted by *Webster's Ninth New Collegiate Dictionary* and the *Random House Dictionary of the English Language* and is assuredly expanding. The word seems to be intensely attractive and scientific-sounding—perhaps because of its basis in mathematics and its wide use in describing activities of the National Aeronautics and Space Administration since at least the days of Sputnik I—and is therefore a strong candidate for ''excessive overuse,'' especially by beginning writers. The eager adoption of the word is often more than a casual happening; in an 11-page manuscript by a young author, *parameter* was used 34 times in the senses here discouraged.

Inasmuch as many knowledgeable readers do not accept the expanded definition (''Some object strongly to the use of *parameter* in these newer senses''— *Random House Dictionary of the English Language*), scientists who remember the importance of transmitting their research findings smoothly and without needless interruption restrain *parameter* to its original meaning in mathematics and statistics. They astutely follow the advice of the *CBE Style Manual* (CBE Style Manual Committee 1983), ''do not use loosely for variable, quantity, quality, determinant, or feature.''

Polysyllabism

Fowler (1965) discussed the "love of the long word" at considerable length, writing in part, "it is a general truth that the short words are not only handier to use, but more powerful in effect; extra syllables reduce, not increase, vigour." He added that a polysyllable should not be avoided if it gives the meaning best, but that "What is here deprecated is the tendency among the ignorant to choose, because it is a polysyllable, the word that gives their meaning no better or even worse."

Elsewhere he noted that "Those who run to long words are mainly the unskilful and tasteless; they confuse pomposity with dignity, flaccidity with ease, and bulk with force. . . ."

The unnecessarily long or stilted words that follow are commonly used in fishery manuscripts where shorter words would often be more effective (the usually appropriate shorter words are in parentheses): *additionally* (*also*), *approximately* (*about*), *currently* or *presently* (*now*), *exacerbate* (*worsen*), *enumerate* (*count, list*), *excepting* (*except*), *exhibit* (*show*), *following* or *subsequent to* (*after*), *individual* (*one, person*), *initially* (*first*), *numerous* (*many*), *partially* (*partly*), *prior to* (*before*), *relationship* (*relation*), *remainder* (*rest*), *respectively* (often avoidable), *retained* (*kept*), *subsequent* (*later*), *succumbed* (*died*), *sufficient* (*enough*), *summation* (*total*), *termination* (*end*), *utilization* (*use*).

principal, principle

Few pairs of words that have no meaning in common are as often confused as *principal* and *principle*. In fishery writing the confusion is hardly random: by far the most often recurring error is the use of *principle* as an adjective where *principal* is the word intended, as in "The principle [make it *principal*] species collected were bluegills, largemouth bass,. . . ." The miscue is assuredly not restricted to fishery writing, but is seen often and sometimes appears in unexpected places; in a book on English usage that goes unnamed here, *principle* is used as an adjective twice, and *principal* once, in nearly identical clauses within eight lines of text. In most fishery manuscripts the problem would be largely solved if *principle* were used only as the noun that it is and *principal* were used only as an adjective. (Although *principal* is in several important senses also a noun, such uses are uncommon in fishery writing.) Learning the details about which word should be used and why is facilitated by the usually same-page proximity of the words in a dictionary.

Punctuation

The use of punctuation—the various standardized marks or signs that clarify meaning and separate structural units in written matter—is fully treated in most grammars and books on usage and style. The succinct, straightforward rules laid down in the *CBE Style Manual* (CBE Style Manual Committee 1983) are especially helpful to fishery scientists. Noted here are selected uses that sometimes weaken or strengthen fishery writing.

• A *colon* preceding a list is most effective if the items in the list are in apposition to an introductory word or words. An example of a sequence seen surprisingly often is, "The collection consisted of seven species of fish. They included: bluegill, walleye,. . . ." Here the colon should follow *fish* and the added

"They included:" should be omitted; it not only weakens the introduction of the list and adds worthless words, but places the colon between the verb and its direct object—a construction generally regarded as a grammatical error.

• The *comma*, frequently used and of course indispensable, is too often misplaced immediately before a parenthesis. Acceptable use is illustrated in the following sentence: "An intense fishery developed for large bluegills, all naturally produced (mean length and weight 212 mm and 285 g), which had become abundant." Here, after the first comma has been set down, the second may neither be omitted nor placed immediately ahead of the opening parenthesis; rather, it advances to its proper place, immediately after the closing parenthesis. (Within sentences in running text, punctuation immediately preceding a parenthesis is rare.)

In nearly all technical writing published in North America, a comma precedes *and* or *or* in a simple series of three or more items, as in "We collected alewives, sea lampreys, and walleyes." Although "Meticulous writing and editing preserves this comma . . ." (Copperud 1980), it is often omitted in general and informal writing. Its consistent use in technical works enhances clarity, comprehension, and readability.

• The *dash* is a useful but often unappreciated punctuation mark that is probably underused in fishery writing—perhaps partly because occasional authorities (e.g., Bernstein 1965) have labeled it as "often overused" and then have given examples of unskillful use. But Barzun (1975) called it the "separating mark par excellence," and Zinsser (1980) wrote that "Somehow this invaluable tool is widely regarded as not quite proper—a bumpkin at the genteel dinner table of good English. But it has full membership and will get you out of many tight corners." Authors who learn the subtleties of its use improve the clarity and readability of their writing.

• An *exclamation point* almost invariably looks (and is) out of place in serious scientific writing. Careful writers who may at first be overenthusiastic about their exciting findings calm down and remove the mark no later than the second draft.

• The *hyphen* draws at least three complaints: (1) Its use at the end of a line of typescript is an open invitation for errors by typesetters and others. (2) It should not ordinarily separate a two-word modifier that includes an adverb ending in *-ly*, as in "A heavily-used trail led to the lake" [omit the hyphen]. (3) Readability is almost equally impaired by the omission of the hyphen between a cardinal number and noun that form a compound modifier as by its insertion between a number and noun that do not form a compound modifier. One should write "It was a 7-day experiment" but "The experiment continued for 7 days." In too many fishery manuscripts, the placement of hyphens between numerals and units of measurement seems to be almost random, unrelated to the function of a combination as a compound modifier.

• *Parentheses* are assuredly useful tools in fishery writing, though they occasionally rise up and strike down readability in unexpected places—as when a literature citation (or other information) in parentheses is referred to in the subsequent open text: "No bluegills of age-group 3 were collected in Mud Lake (Smith 1996). He believed. . . ." Here a reader is at least momentarily taken aback by the unexpected reference in the open text to the parenthetic material that precedes. Readability is improved if the name of the author referred to is brought into the open: "Smith (1996) reported that no bluegills of age-group 3 were collected in Mud Lake. He believed. . . ."

In running text, the juxtaposition of two parenthetic expressions, each with its pair of parentheses "(*Salmo salar*) (Figure 3)," yields a clumsy look and surely disconcerts some readers. The irregularity can usually be avoided by restructuring the sentence, or by separating the two parenthetic elements with a semicolon (less often with a comma or colon) within a single pair of parentheses. The enclosure of one pair of parentheses by another pair, as in "Centrarchids (notably bluegills, largemouth bass (*Micropterus salmoides*), and redear sunfish) . . ." is nearly always confusing or distracting. Although recasting the sentence may be the remedy of choice, the interior parenthesis can often be smoothly set apart with commas or dashes or enclosed in brackets.

• Closing *quotation marks* fall outside commas and periods in conventional American usage.

• In a too often seen version of the backward sentence, an author writes roughshod over a series of *semicolons* and completes the sentence on the other side, as in this sentence about concentration histograms, adapted from a report on polychlorinated biphenyls: "Peak 3, sample 6; peak 4, sample 7; and peaks 5 and 8, sample 10 were considerably higher in sediments than in fish." As in nearly all backward sentences, readability is improved if the last part is put first: "Four peaks were higher in sediments than in fish: peak 3, sample 6; peak 4. . . ."

• The *slant line* (diagonal/slash/solidus/stroke/virgule) has several long-established uses (e.g., as a mathematical sign for division and as a substitute for *per* to show rates or concentrations) that make it invaluable in fishery writing. In recent years, however, its use has greatly expanded; it has been faddishly advanced as a mere word separator (nomenclature/taxonomy/range) or as a lazy replacement for any of a number of words that an author—and especially a reader—might find difficult to identify. A comment by Zinsser (1980) applies to all of these inept usages, "the slant has no place in good English."

quite

The original meaning of *quite*—absolutely, completely, to the fullest extent or degree—was broadened by colloquial use to be a substitute for *somewhat* or *rather*, a usage recognized in dictionaries but not approved by all authorities on writing. In fishery manuscripts, *quite* seems more often to mean "somewhat" or "to some degree" than "absolutely" or "completely," but the meaning cannot always be discerned. One author insisted on quoting "quite abundant" from a description published many years ago, because he could not be certain of the intended meaning. Careful writers who want to ensure that their work is unmistakably understood simply avoid the troublesome *quite*.

relation, relationship

Readability of a manuscript is ultimately impaired by the cumulative effects of minor irritants. One of these is the overuse of a long variant of a word where a shorter one serves at least as well. A frequent example in fishery manuscripts is the use of *relationship* for *relation*, especially in passages dealing with topics such as length–weight relationship, body–scale relationship, predator–prey relationship, and so on, where the word is repeated several to many times within a few paragraphs of text (in a manuscript at hand, *relationship* was repeated eight times on one page). Although some of the repetition commonly reflects weak composition, the unnecessary -ship nevertheless exacts its toll—sometimes small, sometimes significant. Fowler (1965) wrote: "But to affix -ship to *relation* in any of its other, or abstract, senses is against all analogy; the use of -ship is to provide concretes (*friend*, *horseman*, . . .) with corresponding abstracts [*friendship*, *horsemanship*, . . .]; but *relation*, except when it means related person, is already abstract, and one might as well make *connexionship*, *correspondenceship*, or *associationship*, as *relationship* from *relation* in abstract senses." In fishery manuscripts, eliminating one-third of the letters and the long extra syllable by writing *relation* often contributes to smoothness and readability, with no loss of precision or understanding.

reside

"A small number of brook trout reside in Cedar Creek" somehow suggests handsome fish living in opulence, with all the space, tasty food, invulnerable cover, and opportunity for reproduction that any fish could ever want. The sentence somehow overstates the matter of biological interest here, which might better be written, "A few brook trout live in Cedar Creek," or "Cedar Creek supports a few brook trout." Authorities on writing consider *reside* to be pretentious where *live* will do (Copperud 1980), unless (according to Evans and Evans 1957) "it describes the act of living in an important or pretentious residence (*The governor resides at Albany*). . . ." The word thus seems especially out of place in fishery manuscripts in statements such as "Sludge worms reside in the sewage lagoon," or "The parasite resides in the lower intestine." In fact, *reside* always seems out of place when it replaces *live* in referring to any animals other than rich people.

respectively

The inherent attractiveness of the long, multisyllable *respectively* often leads to its insertion where it contributes nothing to clarity or effectiveness. In the sentence, "The mean total lengths at ages 1, 2, and 3 were 20, 30, and 36 cm, respectively," the word serves no useful purpose if one accepts the almost certain likelihood that young fish increase in length as they increase in age. And in various other usages where nothing is respective to anything, the word contributes only syllables and length, and perhaps confusion: "The two smallest fish weighed 4 and 7 g, respectively"; "The mean weight and total length of the sea lampreys were 225 g and 500 mm, respectively"; "Three tagged fish were recaptured after being at large for 15, 37, and 64 days, respectively"; or "Correlation coefficients were 0.815 for steelhead eggs, 0.891 for coho salmon eggs, and 0.797 for chinook salmon eggs, respectively."

In many situations where *respectively* may be technically acceptable, its use stalls all forward motion, as in "Bluegill population estimates in 1983, 1985, and 1987 were 10,600, 12,400, and 25,000, respectively." Here the reader is obliged to stop and back up to make the pairings that a more considerate author could have made far more easily, in fewer total letters and without the awkward juxtaposition of five-digit numbers, by writing "Bluegill population estimates were 10,600 in 1983, 12,400 in 1985, and 25,000 in 1987."

Although the use of *respectively* sometimes facilitates the orderly arrangement of information in a sentence, the word is often overused in fishery writing. Fowler (1965) noted that pretentious writers drag the word in at every opportunity for the "air of thoroughness and precision" that it is supposed to give to a sentence, and suggested that 9 of 10 sentences in which *respectively* appears would be improved by its removal; and Follett (1966) wrote that the word probably betrays "a widely felt fondness for polysyllables that make one's meaning sound more technical than it is." Careful writers weigh each use of *respectively* to ensure that the sentence in which it appears cannot be written more clearly, directly, and pleasingly without the long extra word.

Spelling

Although poor spelling may not be correlated with intelligence, the ignorance or carelessness that it reflects assuredly does not flatter a reader's image of the writer. I believe that Barzun and Graff (1977) had it right: "Spelling is one of the decencies of civilized communication. No doubt there are reasons, physical and psychological, for misspellings, but a reason is not an excuse. This is especially true if you are fond of treacherous words such as 'corollary' and 'supersede' and 'apophthegm'; the least you can do is to learn to spell them. It is like paying for what you use instead of stealing and spoiling it. As to the commoner words, such as 'receive,' 'benefited,' 'indispensable,' and 'consensus,' it should be a matter of pride with you not to exhibit your work pockmarked with infantile errors."

The following words must be included in any list of common misspellings in fishery manuscripts (the most common incorrect spellings are in parentheses; "misspellings" followed by an asterisk are acceptable variants in some quarters, but not in fishery writing in North America): *aberrant* (abberant), *accommodate* (accomodate), *benefiting* (benefitting*), *buoy* (bouy), *consensus* (concensus), *desiccate* (dessicate), *desirable* (desireable), *en route* (enroute), *fluorescent* (florescent, flourescent, flurescent), *gray* (grey*), *iridescent* (irridescent), *led* (lead), *liquefy* (liquify*), *Notemigonus crysoleucas* (*N.* chrysoleucas), *occasion* (occassion), *occurring* (occuring), *precede* (preceed), *separate* (seperate), *supersede* (supercede*), *under way* (underway).

Perhaps the most common deviation from currently conventional spelling seen in fishery manuscripts is the use of the single word *underway* in place of the two-word adverbial phrase *under way*. Despite the rather common use of the single-word form for many years, dictionaries have not yet given it even grudging acceptance. Fishery authors can feel free to write *under way* with impunity, for the one-word adjectival form (e.g., "underway refueling") is almost nonexistent in fishery writing.

Scientific names of animals and plants and geographic place names (often written from memory, one suspects—or with scant care) are commonly misspelled.

Addendum. Although the matter is not strictly one of spelling, it is allied closely enough to be stated here: In the formal language of fishery reports for technical audiences, contractions (*can't, don't, it's, we've,* . . .) almost never fail to loudly call attention to their inappropriateness. They may sometimes be acceptable in informal reports.

Split Infinitives and Compound Verbs

Perhaps every authority on English usage has commented on the split infinitive (the interposition of a word or words between *to* and the infinitive form of the verb, as in "to better understand the technique" or "to easily increase the scope of the work"). All or nearly all authorities agree that the two- or three-century-long, grammatically insupportable taboo on the construction is now gone. Early on, Fowler (1926) considered a split infinitive to be "preferable . . . to real ambiguity, & to patent artificiality," and Curme (1931) believed that "the insertion of the adverb . . . between *to* and the infinitive cannot even in the strictest scientific sense be considered ungrammatical." Follett (1966) called it "the one fault that everybody has heard about and makes a great virtue of avoiding and reproving in others," but agreed that "the split infinitive has its place in good composition." And Copperud (1980), summarizing the consensus of seven critics, wrote that "infinitives may be split when splitting makes the sentence read more smoothly and does not cause awkwardness."

The construction to be usually avoided because it is still frowned on by many authorities is the insertion of more than one or perhaps two words between *to* and the infinitive. Most such multiple-word splits are automatically ruled out because they clearly "cause awkwardness." It is unlikely, for example, that a statement such as "We did not fail in our appointed task to, under all conditions of snow, rain, and gloom of night, collect the samples at the scheduled times" would ever be considered acceptable.

A moderate relaxation of the fear of splitting infinitives should help fishery scientists overcome a tendency to try to avoid a split infinitive where no infinitive exists. Writers sometimes obviously hesitate to insert an adverb between the elements of a compound verb (as in "they have often used this technique"), even when that is clearly the most effective place for the word. Bernstein (1965) wrote that "some writers, blinded by the split-infinitive obsession, seem determined not to split anything except hairs. Thus, they will not permit an adverb to divide elements of a compound verb. There is no reasonable foundation for this attitude. The truth is that more often than not the proper and natural place is between the parts of a compound verb." Copperud (1980) expressed a similar view in offering the consensus of five authorities: "An erroneous idea that compound verbs . . . should not be separated by adverbs (*have* easily *seen*, *will* soon *go*) has become widespread. . . . Most commentators agree that it is apparently an offshoot of the prohibition against splitting an infinitive, which . . . is also a superstition."

Stacked Modifiers

Clarity and soundness of expression are often gravely impaired by the indiscriminate and increasingly persistent stacking of nouns and adjectives that collectively modify a terminal noun, as in "the floating barge sport fish catch statistics. . . ." Multisyllable words are usually preferred, and the terminal nouns are often far removed: "The extracts were tested for mutagenicity in the Ames, rat hepatocyte unscheduled DNA synthesis (UDS), and Chinese hamster ovary hypoxanthine-guanine phosphoribosyl transferase (CHO/HGPRT) assays." Such combinations make for easy writing, since little thought about composition is required while successive modifiers are being added to the stack, but they make for fatiguing and difficult reading. Those to whom English is native and to whom the general subject is familiar can perhaps suppress their distaste and discern the intended meaning. But others, such as foreigners whose grasp of the language is limited, are likely to be completely lost. All readers would be grateful if, in the sentence just quoted, the last word were brought nearer to the head of the sentence: "The extracts were tested for mutagenicity in three assays: Ames, rat hepatocyte . . ., and Chinese hamster. . . ."

Straining to be Scholarly

On occasion—too often, perhaps—an author (usually young) tries in a single paper to become established as one of the world's Great Scholars. The materials vary but the symptoms are the same: the rhetoric and syntax are pompous, flatulent, and obscure; technical words abound; no one-syllable word appears where a multisyllable equivalent can be fitted in; and ideas and concepts are twisted into formidable complexities. Amidst all this scholarly elegance, oddly enough, many vogue words and cliches are likely to appear. Mathematical and statistical treatments are eagerly paraded, and recently learned formulas and equations are likely to be supplied for routine statistical computations. All the details are inflicted on the reader in the text and in tables of pure statistical data that effectively obscure biological findings.

A discerning, sophisticated reader is not impressed by a paper in which the simple is made to seem difficult and ideas are buried in a mass of verbiage. But if the difficult is handled simply, in straightforward English, the reader knows that the author has made a scholarly treatment.

that, *which*

The distinction between *that* for the introduction of restrictive clauses ("This is the fish that Jack caught") and *which* for nonrestrictive clauses ("This fish, which Jack caught, is a bluegill") was deemed worth observing by all of the authorities surveyed by Copperud (1980) except Evans and Evans (1957), who termed the distinction "an invention of the grammarians." They further noted that "In actual practice, *which* is not often used in a defining [restrictive] clause today. . . ." But Bernstein (1977) was much nearer the mark when he wrote, "Rarely do people use *that* when the word should be *which*, but quite often they use *which* when the better word would be *that*"—the usage that often weakens sentences in fishery manuscripts. The distinction between *that* and *which* is not always easy to make; however, it usually yields to the test of commas: If the clause cries out to be set off with commas or ungrudgingly yields to their insertion ("These three fish, which had been injured, were found dead"), it is nonrestrictive; the relative pronoun needed is *which*, and in scientific writing the insertion of commas adds immediate clarity. Typically, if the clause cannot smoothly be set off with commas ("The three fish that were injured soon died"), it is restrictive, and *that* is the word of choice. *Which* is often used in this situation because it "tends to sound more formal and they [the writers] think they are being more elegant" (Bernstein 1977). Two technically errant *which*'s from a fishery manuscript further illustrate the point: "The fish stocked were from a hatchery strain which [make it *that*] had been selected for fast growth," and "Catchability is an important aspect of fish behavior which [make it *that*] has frequently been overlooked."

Fowler (1965) wrote that "there would be much gain both in lucidity and in ease" if the distinction between the two words were maintained—to which perceptive readers of fishery literature might add, "in strength and in effectiveness." These gains are left unclaimed and weakness is substituted when *which* is used where *that* is the sounder choice.

use, using, utilize, utilization

The verbs *use* and *utilize*, insofar as one can determine from any standard dictionary, are synonyms. Yet many fishery writers seem able to write only *utilize*, or circumvent the verb in favor of the insidiously more attractive five-syllable, 11-letter *utilization*. Writing would read better if loyalty to the longer words were not so strong.

Perhaps more important than this misplaced loyalty to the long word is the distressingly frequent appearance of *using* in scientific writing in situations where it has no tenable subject, as in "Production was estimated using a creel census," where the only subject available—production—assuredly cannot use a creel census. Nevertheless, the usage has become so deeply entrenched that the present dissent is subject to being discarded as a mere curmudgeonly crotchet.

In the eyes of others, almost all participles—dangling or not—are acceptable because they were "used by Chaucer and all our great writers before or since" (Evans and Evans 1957); or *using* may have by now joined the ranks of grammatically independent participles that have become sanctioned by long use and abuse (McCartney 1953); or the usage has simply been declared unobjectionable (CBE Committee on Graduate Training in Scientific Writing 1968). Nevertheless, it is contended here that participles (in other than absolute constructions) that are not anchored to tenable subjects are often distracting to some readers, and misattached ones nearly always are. No gold star for elegance can be claimed by one who writes, "The trout were caught using a dip net" or "The eagles were observed using binoculars." Respectability is conferred on each sentence when *using* is merely replaced with *with*, or when the misleading passive voice is replaced by the active: "We used a dip net to catch the trout," or "We watched the eagles through binoculars."

Weasel Words

Frequently the data in hand are insufficient to permit unqualified conclusions or the laying down of sharply defined principles. These same data may nevertheless contribute significantly to knowledge and permit real advancement of understanding. The authors who know the data best and have analyzed them meticulously are obligated to offer the best interpretation that can be mustered. Doing so may require that the remarks be hedged with appropriate reservations. If authors do this skillfully, they effect a clear presentation of data and a convincing statement of views, without laying themselves open to the accusation of being dogmatic in their opinions or to the danger of being flayed by those who amass additional evidence that proves some of the views to be wrong.

All of these precautions against erroneous interpretations of uncertain data are commendable; however, avoiding the positive statement can become a habit that results in an overuse of "weasel words." Some authors shy away from a definite conclusion when no other is possible. When they fit a regression to a set of points and offer a graph that shows the points to cluster tightly about the derived curve, they write that the curve "seems to" fit the data. If they have watched small fish in a stream season after season and have never seen one swimming against the current, they observe that the fish "appear to" move with the current. Or they may reinforce the hedges by writing that the data seem to suggest that the fish may move with the current. Such evasions of the obvious take many and sometimes subtle forms. They do not necessarily destroy a paper, but certainly weaken it. Readers, unconsciously perhaps, are less convinced by the opinions of authors who are never sure of themselves.

Overly conservative weasel-wording has introduced usages that have almost become established jargon. Two phrases that come to mind are "compare favorably with" and "associated with." If *a* and *b* are closely similar, why not say they are, rather than observe that *a* "is comparable to" or "compares favorably with" *b*? Seemingly almost anything can be "compared" with something else, and *favorably* has not been given a useful technical definition. Again, why say that the rapid eutrophication of a lake has been "associated with" an increased discharge of domestic sewage and more intensive fertilization of farmlands, when the evidence is conclusive that the sewage and fertilizer have been the cause?

while

"In general, the writer will do well to use *while* only with strict literalness, in the sense of *during the time that*" (Strunk and White 1959). This excellent advice applies particularly to scientific writing, where comprehension should be immediate and unmistakable. Unfortunately, however, relatively few fishery writers accept Strunk and White's admonition; rather they use *while* where *although*, *and*, *but*, *whereas*, or a semicolon would serve better. Bryant (1962) wrote that "The connective *while*, meaning "whereas," occurs as frequently in standard English as *whereas* itself"—and this relation assuredly holds in fishery manuscripts. The use of such a shifty word, especially in its trickiest position at the beginning of a sentence, ensures ambiguity or imprecision, either briefly, "While the tractor moved ahead, it dragged the seine through the pond" (where context saves the day), or forever, "While we set trap nets in the tributary the other crew trawled in the bay."

All in all, it seems clear that *while*, at least at the head of a sentence or clause, could without loss be relegated with *former* and *latter* to the lexical dustbin of the fishery writer. However, it can sometimes be used effectively in the interior of a sentence in sequences where its use in the temporal sense cannot be even momentarily misconstrued: "In an aquarium, the turtles surfaced every 3 minutes while actively swimming and every 12 minutes while resting."

with

Some words serve comfortably as different parts of speech, but *with* often weakens sentences in which it is used as a conjunction of sorts, rather than as the preposition to which its use is restricted by careful writers. In its most offensive (and perhaps most frequently questionable) use in fishery manuscripts, it ineffectively connects unrelated elements of sentences, usually in the form of anchorless, independent addenda tacked to essentially completed sentences: "with the largest fish being 80 cm long and weighing 8 kg," or "Station temperatures were taken every 2 hours with water samples collected every 8 hours," or "single-pair matings are recommended with family lots held separately until hatching." Sometimes the awkward usages can be avoided only by recasting the sentence; in other situations *with* can be replaced by *and* or a semicolon.

Acknowledgments

I am deeply grateful for having had a pleasant and productive association with my mentor and long-time friend, Ralph Hile, who introduced me to editing and nurtured my interest. Many of the discussions of problems included here are based on *Editorial Memos* that he and I prepared in 1965–1969. For reading the manuscript and making suggestions, I thank R. W. Gregory, C. H. Halvorson, D. K. Harris, P. A. Opler, and N. M. Wells of the U.S. Fish and Wildlife Service Office of Information Transfer; W. N. Eschmeyer, California Academy of Sciences, San Francisco; R. L. Kendall, American Fisheries Society, Bethesda, Maryland; V. S. Kennedy, Horn Point Environmental Laboratories, University of Maryland, Cambridge; C. J. Sinderman, National Marine Fisheries Service, Northeast Fisheries Center, Oxford, Maryland; and L. Wells, National Fisheries Research Center–Great Lakes, Ann Arbor, Michigan.

References

American Heritage Dictionary of the English Language. 1969. American Heritage Publishing, New York, and Houghton Mifflin, Boston.
Barzun, J. 1975. Simple & direct: a rhetoric for writers. Harper & Row, New York.
Barzun, J., and H. F. Graff. 1977. The modern researcher, 3rd edition. Harcourt Brace Jovanovich, New York.
Bernstein, T. M. 1965. The careful writer: a modern guide to English usage. Atheneum, New York.
Bernstein, T. M. 1977. Dos, don'ts & maybes of English usage. Times Books, New York.
Bryant, M. M., editor. 1962. Current American usage. Funk & Wagnalls, New York.
CBE Committee on Graduate Training in Scientific Writing. 1968. Scientific writing for graduate students: a manual on the teaching of scientific writing. Council of Biology Editors, Bethesda, Maryland. (Reprinted 1983.)
CBE Style Manual Committee. 1983. CBE style manual: a guide for authors, editors, and publishers in the biological sciences, 5th edition. Council of Biology Editors, Bethesda, Maryland.
Copperud, R. H. 1964. A dictionary of usage and style. Hawthorn, New York.
Copperud, R. H. 1980. American usage and style: the consensus. Van Nostrand Reinhold, New York.
Curme, G. O. 1931. Syntax. Heath, Boston.
Evans, B., and C. Evans. 1957. A dictionary of contemporary American usage. Random House, New York.
Follett, W. 1966. Modern American usage: a guide. (Edited by J. Barzun and others.) Hill & Wang, New York.
Fowler, C. A. 1935. Employer's Mutual Insurance Company vs. Tollefsen. Wisconsin Reports 219:434. (Not seen; from correspondence.)
Fowler, H. W. 1926. A dictionary of modern English usage. Clarendon Press, Oxford, England.
Fowler, H. W. 1965. A dictionary of modern English usage, 2nd edition. (Revised by E. Gowers.) Oxford University Press, New York.

Hile, R. 1971. No read—no write. Transactions of the American Fisheries Society 100:394-395.

Kilpatrick, J. J. 1984. The writer's art. Andrews, McMeel & Parker, Kansas City, Kansas.

McCartney, E. S. 1953. Recurrent maladies in scholarly writing. University of Michigan Press, Ann Arbor, Michigan.

Morris, W., and M. Morris. 1985. Harper dictionary of contemporary usage, 2nd edition. Harper & Row, New York.

Oxford English Dictionary (The Compact Edition of the Oxford English Dictionary). 1971. Oxford University Press, New York.

Random House Dictionary of the English Language, 2nd edition. Unabridged. 1987. Random House, New York.

Robins, C. R., R. M. Bailey, C. E. Bond, J. R. Brooker, E. A. Lachner, R. N. Lea, and W. B. Scott. 1980. A list of common and scientific names of fishes from the United States and Canada, 4th edition. American Fisheries Society Special Publication 12.

Robins, C. R., R. M. Bailey, C. E. Bond, J. R. Brooker, E. A. Lachner, R. N. Lea, and W. B. Scott. In press. Common and scientific names of fishes from the United States and Canada, 5th edition. American Fisheries Society Special Publication 20.

Strunk, W., Jr., and E. B. White. 1959. The elements of style. Macmillan, New York.

Webster's Ninth New Collegiate Dictionary. 1988. Merriam-Webster, Springfield, Massachusetts.

Webster's Third New International Dictionary of the English Language Unabridged. 1986. Merriam-Webster, Springfield, Massachusetts.

Zinsser, W. 1980. On writing well: an informal guide to writing nonfiction, 2nd edition. Harper & Row, New York.

Problems with Gray Literature in Fishery Science

BRUCE B. COLLETTE

National Marine Fisheries Service
Systematics Laboratory
National Museum of Natural History
Washington, D.C. 20560, USA

Abstract.—Fishery scientists and their organizations are responsible for posing important questions, gathering relevant data, analyzing these data, and making the results available to the fishery community. This process is hindered by the production of gray literature: that is, written information that is produced and distributed without adequate review. Gray literature takes time and effort to produce, although not as much as manuscripts for peer-reviewed journals. Because it is poorly evaluated, it lacks credibility. Authors of gray literature reports feel that they have done their job by writing the report but they have not completed the necessary tasks of producing credible information and ensuring that the information is distributed in a readily available medium to those who need it. Gray literature is hard to retrieve because it is usually not abstracted or well distributed. Some gray literature is produced because of contractual demands and may serve as data archives if made available from national repositories or the institution producing the reports. Early release of scientific information in the gray literature may jeopardize subsequent publication in the formal literature. Fishery agencies and fishery scientists should avoid producing gray literature and concentrate their efforts on producing good papers that will be accepted and published in the permanent scientific literature.

It is difficult to define gray literature, valid publication, or scientific publication. The following, necessarily long definition was put together by the Council of Biology Editors (Day 1983). "An acceptable primary scientific publication must be the first disclosure containing sufficient information to enable peers (1) to assess observations, (2) to repeat experiments, and (3) to evaluate intellectual processes; moreover, it must be susceptible to sensory perception, essentially permanent, available to the scientific community without restriction, and available for regular screening by one or more of the major recognized secondary services . . . ." This definition means publication in a journal or other source document readily available within the scientific community. Further, "peers of the author" is accepted as meaning prepublication peer review (Day 1983). Scientists usually use the term "gray literature" to refer to literature produced without peer review. Librarians use a definition based only on availability, "material not available through normal bookselling channels" (Vickers and Wood 1982; Hesselager 1984; Wood 1984). Much gray literature is produced by fishery organizations and private contractors to those organizations. The lack of peer review raises questions about credibility of the information contained therein and accounts for the frequent warnings not to cite such information without permission of the author. I am concerned about the lack of peer review of such material, its relative unavailability, and the effects that producing gray literature have on fishery scientists and the organizations for which they work.

Before discussing specific problems, I will outline hierarchies of information output in the National Marine Fisheries Service (NMFS) as an example of the situation in other fishery organizations. Basically, NMFS has three publication outlets for scientific papers plus several types of gray literature. These differ in the degree of rigor applied in the review and editorial process.

(1) Peer-reviewed journals. *Fishery Bulletin* is a quarterly, peer-reviewed journal open to suitable papers from authors from any organization. *Marine Fisheries Review,* also peer-reviewed, is aimed at a more general audience.

(2) Irregular peer-reviewed series. *NOAA* [National Oceanographic and Atmospheric Administration] *Technical Report NMFS,* combining two previous series (*Special Scientific Report Fisheries* and *Circular*), accommodates papers that are inappropriate for *Fishery Bulletin* because of their great length or limited scope.

(3) Irregularly issued and variously edited reports. *NOAA Technical Memorandum NMFS-F-* comprises series issued by each of the fishery centers and regions. These series provide a relatively quick outlet for documents for which formal review and complete editorial processing are not appropriate or feasible due to time constraints. Documents in this series reflect sound professional work that can be referenced in formal journals. Copies of *Technical Memorandums* can be purchased from the National Technical Information Service, Springfield, Virginia, USA.

(4) Gray literature. Each of the fishery centers within NMFS and some of the individual laboratories administered by those centers issue various kinds of technical reports and discussion documents. These have limited distribution and are not peer-reviewed. The Woods Hole Laboratory Reference Series, Middle Atlantic Coastal Fisheries Center Technical Publications, Sandy Hook Laboratory Technical Series Reports, Northwest Center Processed Reports, and Southwest Center Administrative Reports are but a few examples. Discussion documents are also produced for the fishery management councils.

The Canadian Department of Fisheries and Oceans has a similar hierarchy. The *Canadian Journal of Fisheries and Aquatic Sciences* is one of the premier fishery journals in the world. Less regular but still citable are *Canadian Technical Report of Fisheries and Aquatic Sciences* and some analogous series. *Canadian Data Report of Fisheries and Aquatic Sciences* is not citable "without prior written clearance from the issuing establishment."

Similarly, the former International Commission for the Northwest Atlantic Fisheries (ICNAF) published a *Research Bulletin* as well as *Selected Papers*. Its successor, the Northwest Atlantic Fisheries Organization (NAFO) publishes the *Journal of Northwest Atlantic Fishery Science* and *Scientific Council Studies*. But both ICNAF and NAFO, as well as ICES (International Council for the Exploration of the Seas), also produce "Working Documents" for use at the meetings. These are not necessarily peer reviewed and bear the warning that they should not be cited without permission of the author.

Another example concerns the two international tuna commissions. The Inter-American Tropical Tuna Commission (IATTC) produces a peer-reviewed *Bulletin,* as well as an *Annual Report* and irregular *Special Reports* that are sent to a much more limited audience. The International Commission for the Conservation of Atlantic Tunas (ICCAT) publishes a biennial *Report*, a *Statistical Bulletin*, and a *Data Record*. But ICCAT also issues an annual *Collective Volume of Scientific Papers* for use at the meetings of the Standing Committee on Research and Statistics. A cautionary note follows the cover page: "In order to ascertain the validity of the data and the conclusions expressed in each document, users are requested to contact authors. Some of the documents are not citable without prior consultation with the authors."

Many states and provinces produce their own gray literature such as project reports based on funding received from the U.S. Environmental Protection Agency, the U.S. Fish and Wildlife Service, etc. In addition, private environmental consulting groups produce hundreds of reports annually; many of these have very restricted distributions because they contain proprietary information.

The gray literature problem is going to become greater with the advent of desk-top publishing. Why take the time and trouble of sending a manuscript to a journal, trying to revise the manuscript, hoping to have it accepted, and then waiting months or even years for publication when one can just sit down at a personal computer, enter the document, and issue it all on the same day? The old statement about computer programming—garbage in, garbage out—will apply to many "self-published" documents. Scientific papers, however, need credibility.

In addition to the problem of credibility, gray literature poses other major problems. Much gray literature is not easily available. It frequently is neither found in libraries nor listed by abstract services. It may be obtainable only from the author. This is a difference between gray literature and informal technical series such as the *NOAA Technical Memorandums*, which are available through the National Technical Information Service. When authors cite gray literature (frequently their own) in journals, they create a demand for data that is hard to satisfy. If these data are obtainable only from the author, it is very difficult for journal readers to validate the author's statements about them. When that author retires or dies, is usually becomes impossible to find the data. This is the prime reason why most editors refuse to list gray literature in bibliographic sections of papers submitted for publication in their journals. Gray literature is usually referred to in a footnote or parenthetical statement in the same fashion as a personal communication.

Production of gray literature may jeopardize subsequent publication in formal journals. It is accepted scientific practice to publish papers once and only once. Some scientists have found that reports required for contracts or grants were considered by some editors to be published and available, and that a formal paper based on the same data was not acceptable for publication a second time in a peer-reviewed journal.

A related problem concerns promotions of fishery scientists. When they evaluate scientists for promotion, most universities and many fishery agencies concentrate on the papers the scientists produce in peer-reviewed journals but give little or no credence to gray literature. Thus, someone with a long list of gray literature reports may be at a competitive disadvantage compared to someone who has published similar material in peer-reviewed journals.

Perhaps the most serious criticism of gray literature is the effect it produces on the author and his or her organization. Authors tend to feel that their job is completed once they have generated some sort of report. But if the information is not fully credible and if it is not easily available, the author and the organization have not completely fulfilled their responsibilities. Most fishery data are produced within organizations that derive part or all of their financial support from some governmental unit. The public has a right to expect that data whose acquisition and interpretation were publicly financed are available to fishery managers. Authors and administrators have a clear responsibility to see that information is adequately evaluated and available to the appropriate user groups. In most instances, these goals are met by publication in an established peer-reviewed journal.

If there are so many problems with gray literature, why is there so much of it? One answer is that it fulfills a need and is comparatively easy to produce; some types of information need to reach the user rapidly and any kind of peer review takes time. However, I question why some of this information could not simply be transmitted in letters or memoranda without production of formal documents. A second argument is that it takes time to gather together all the information found in a gray literature document; therefore, the author deserves "credit" for producing it. Some fishery scientists have this attitude and include gray literature in their curriculum vitae. However, they do not list all the memos and letters they have written, some of which also took much time and effort. Why not exert a little more effort and produce a manuscript worth submitting to a journal? A third argument is that there is training value for young scientists in producing gray literature. My response is that if a paper is not subjected to peer review, the process is incomplete and so is the so-called "training." I believe that a fourth reason for producing gray literature is to avoid review and possible criticism from outside a working group or laboratory. There is an attitude among some laboratory personnel that they possess all of the expertise on a particular topic. This is dangerous for the scientist, the discipline, and the organization.

Recommendations.—I recognize that some gray literature must be produced, but I urge that its production be minimized. This requires that the distinctions between gray and peer-reviewed literature be clearly understood. Fishery scientists and fishery agencies should concentrate their efforts on producing good papers that will be reviewed, accepted, and published to become part of the permanent scientific literature. Managers should encourage fishery scientists to publish worthwhile papers in established journals. They should discourage production of gray literature by refusing to approve release of information in this manner unless there is specific reason for doing so. Gray literature that is produced as the result of contractual demands should be made available through a national repository or the institution producing the reports. Funding agencies and supervisors could require that final reports be formatted for submission to peer-reviewed journals. When gray literature such as annual stock assessments must be produced for a user group, scientists should be strongly encouraged to summarize these assessments periodically to ensure that the conclusions and most important supporting data are incorporated into the permanent scientific literature. Fishery scientists suffer by not having their manuscripts peer-reviewed, and they should be aware that early release of scientific information in the gray literature may jeopardize subsequent publication in the formal literature.

Acknowledgments

I thank John Hunter for asking me for my opinions on gray literature. My colleagues at the Northeast Fisheries Center, National Marine Fisheries Service helped focus my ideas during a staff meeting at which the abstract of this paper was discussed. Drafts of the manuscript were read and improved by Thomas A. Munroe, Michael Vecchione, and Austin B. Williams of the Systematics Laboratory and Daniel M. Cohen of the Los Angeles County Museum of Natural History.

References

Day, R. A. 1983. How to write and publish a scientific paper, 2nd edition. ISI Press, Philadelphia.

Hesselager, L. 1984. Fringe or grey literature in the national library: on "papyrolatry" and the growing similarity between the materials in libraries and archives. American Archivist 47:255–270.

Vickers, S., and D. N. Wood. 1982. Improving the availability of grey literature. Interlending Review 10(4):125–130.

Wood, D. N. 1984. The collection, bibliographic control and accessibility of grey literature. International Federation of Library Associations 10:278–282.

Writing for Fishery Journals
© Copyright by the American Fisheries Society 1990

Graphic and Tabular Display of Fishery Data[1]

Victor S. Kennedy and Deborah C. Kennedy

Horn Point Environmental Laboratories
University of Maryland
Box 775, Cambridge, Maryland 21613, USA

Abstract.—Most scientific writing involves analysis, interpretation, and reporting of quantitative data. Such writing must be literate and numerate to attract and hold the attention of busy readers, and careful crafting of figures and tables enhances this task. Ideally, figures preserve detailed information for study and encourage reasoning about that information. They allow a large body of data to be encoded in a compact space. Studies have revealed how graphic-perception tasks are performed when one decodes such information. For example, it is easier to judge the position of points along a common scale than along identical but nonaligned scales. In turn, both of these judgments are more accurate than a judgment (in order of decreasing accuracy) of length, angle, slope, area, or volume. This knowledge has led to the development of techniques for graphing data with visual clarity. We use fisheries and other data to illustrate some of these graphic techniques. Like figures, tables may be used to present raw data, summaries of those data, or analyses based on them. We show ways to present such tabular material clearly and comprehensively.

Scientists should strive to be literate, articulate, and numerate, and should become familiar with the best ways to illustrate their writing with figures and tables. Literacy and articulacy (Gowers 1977) ideally are cultivated during high school and college; numeracy develops especially in graduate school. However, a scientist's education in graphic illustration is usually self-education and imitative of what others have done. We hope here to foster an understanding of graphic and tabular methods of enhancing the presentation and analysis of fisheries data. We present a brief overview of some goals of graphic display and the development of graphic methods, and follow with general principles and standards of graphic design. We then discuss our survey of figures and tables in four fisheries journals and conclude with suggestions and new methods for the display of data.

Graphic Display

Figures and tables should be structured toward at least two ends. The first is to order related data into a logical sequence, perhaps according to time (the most common construct), or magnitude (e.g., highest to lowest), or pattern of observation or experimentation. The second is to place one or more variables or data sets in juxtaposition so that their changes are seen in relation to one another, the

---

[1]Dedicated to those major professors who ensure that their students' theses and dissertations are pruned and shaped into succinct acceptable manuscripts *before* they are sent to journal editors.

total, or both. Such ordered relationships are aids to organization; the goals are to enhance understanding of the data, to expedite the drawing of conclusions or the making of decisions, and perhaps to persuade.

Gowers (1977) noted that writing is an instrument for conveying ideas from one mind to another. Similarly, figures aid the flow of information that occurs when data are recorded, encoded in a message, and transferred to and decoded by an observer. Well-designed figures illuminate relationships among ideas or data that are difficult to explain precisely in words. They improve understanding, draw attention to major issues, and enhance retention of the message. They also save the time of the observer because extraneous or trivial material is dispensed with, permitting close study of the remaining essentials for trends or relationships— including those once hidden by "noise." Many scientific figures (perhaps a majority) represent time series, although time is not a good explanatory variable (Tufte 1983), but about 40% of scientific figures are relational (i.e., they display two or more variables of a kind other than time, latitude, or longitude). This encourages the viewer to search for possible causal relationships between the variables (Tufte 1983).

In fisheries papers, figures may include diagrams and schematics, illustrations, maps, photographs and photomicrographs, and graphs. They are used
- to display and describe data effectively and without distortion,
- to draw attention to key aspects of the data,
- to place these key aspects into context,
- to analyze and display statistical relationships,
- to uncover relationships that might overwise be hidden or unclear,
- to illustrate nonvisual concepts visually, and
- to speed communication and save the reader's time.

Good figures are accurate, simple and uniform, clear and unambiguous, of pleasing appearance, and structurally well designed in that all design elements are interdependent (Schmid 1983). They are crafted to attract and hold attention. In order to accomplish the goal of clear, rapid communication, one must understand not only the data and their implications, but also the purpose of the basic format. This will allow the format to emphasize the data. It follows, then, that the careful fisheries scientist discards unnecessary detail and employs appropriate design and scaling of graphic material to display that material effectively.

Graphic Development

Although earlier primitive figures are known (Tufte 1983), credit for the first extensive development and use of graphs (dealing with time series in economics) is given to William Playfair of London. He used them in his publications as early as 1786 (Tufte 1983), apparently preferring them to tables for their ability to present the shape of the data and to enable visual comparisons to be made. He also developed the first bar graph (Tufte 1983). Over the ensuing two centuries, modest progress was made in refining graphic styles and methods. However, some researchers have begun recently to use knowledge of graphic perception (the ability of an observer to decode quantitative information encoded on graphs) to explore methods of improving figures. Thus, W. S. Cleveland and his colleagues at AT&T Bell Laboratories have examined elementary tasks of graphic perception performed when one decodes figures visually.

To determine the current "state of the art," Cleveland (1984) examined 377 figures in Volume 207 of *Science* (1980). Thirty percent of the figures had at least one of four problems: 15% lacked an explanation of something on the figure; 10% had items (e.g., symbols) drawn on the figure that were hard to distinguish; 6% were constructed or labeled incorrectly; and 6% had reproduced poorly when printed. Of greater consequence, many figures had an inadequate graphic form, or (worse yet) were based on poorly chosen quantitative information.

Cleveland (1985) sought to find ways to improve this situation. It is known that visual decoding of information involves "preattentive vision," or the immediate comprehension of visual material with no apparent mental effort (Cleveland and McGill 1985). This makes figures different from and often more powerful than tables (discussed later) because preattentive vision grasps geometric patterns of position and estimates magnitudes. Cognitive tasks such as reading labels and scales are also performed when information is decoded, but preattentive vision was singled out for examination by Cleveland and McGill (1985). They used both theoretical and experimental approaches to discover that the graphic perception tasks performed can be ranked in order of accuracy, and a framework for data display can be established by attention to that ranking. Graphic elements (position, length, etc.) become more difficult to judge in the following sequence:

(a) positions along a common scale;
(b) positions on identical scales that are not aligned with one another (i.e., that are offset vertically or horizontally);
(c) lengths, whether arranged vertically or horizontally;
(d) angles or slopes of lines;
(e) area, as in circle or pie graphs;
(f) volume, density, color saturation, or color hue (not commonly used in fisheries graphs).

Figure construction should encode data so they can be decoded as easily as possible. Thus, designs involving estimation of area should be replaced if possible by those involving judgments of position or length. Cleveland (1985) elaborated extensively on these findings, and Cleveland and McGill (1987) provided additional experimental data. For the fisheries scientist with an interest in the subject but limited time to explore it, Cleveland and McGill (1985) is the most accessible and succinct of these reports. We later consider the application of this knowledge of graphic perception to the display of fisheries data. We now offer some generalities about graphic standards and designs, some of which—for example, those about legends and rigorous proofreading—can also apply to preparation of tables.

Graphic Standards and Design

Figures enable large amounts of data to be presented in small spaces; data-poor figures should be replaced by a table, or the data should be reported in the text. Tufte (1983) and Cleveland (1985) showed examples of clearly understandable graphs with more than 1,000 data points that encoded two numbers each. This quality of illustration must be prepared meticulously. The process of graphing data is often iterative; one should not necessarily be satisfied with the first idea that comes to mind when dealing with complex data. The goal should be one of clarity in both data portrayal and explanation of graphic elements.

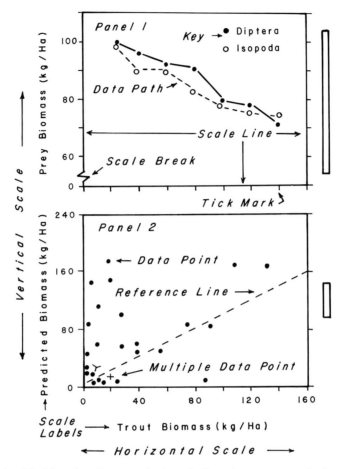

FIGURE 1.—Models of a line graph (panel 1) and scatter graph (panel 2), with identifications of component parts. Vertical rectangles to the right of the panels have the same numerical (scale) dimension. Ha = hectare.

Tufte (1983) urged maximization of the data:ink ratio ("the proportion of a graphic's ink devoted to the non-redundant display of data. . .") and minimization of ink used for nondata features (e.g., borders, axes, grids, half of the lines that form bars and columns). He redesigned the range-bar, box-plot, histogram, and scatter graphs, reducing the number of lines needed for each. However, we have chosen to describe the more usual styles and formats and to emphasize their clear display. We do note that some recent fisheries graphs are "minimalist" (an example is given later) and we expect more to appear with time as attention to graphic display continues.

In terms of shape, the most common is that of a rectangle bounded by solid lines (Figure 1). Sometimes the right and upper boundary lines are not used. However, Cleveland (1985) urged the use of four border lines, each with scales, to provide a defined region that can be searched for data points and to allow easier assessment of points that would be far removed from a scale line if the right and upper borders were missing.

The American Society of Mechanical Engineers has developed standards for graphic display. One of its standards states that oblong horizontal graphs (with vertical:horizontal dimensions in ratios of, for example, 2:3 or 3:4) are preferable to square graphs (ANSC 1979). Among the reasons for this, the eye is more accustomed to reading deviations from the horizontal, it is easier to write and read English labels on a field wider than high, and a wide horizontal display does more justice than a vertical format to sharply varying curves (Tufte 1983). At all costs, however, one must not produce misleading figures. Manipulation of the two scales of a line graph can change the message of the data, even though the data themselves do not change. For example, when vertical and horizontal scales are equal, the data region (the area within the rectangle) is square, it exaggerates in neither direction, and the display is neutral. If the vertical scale is exaggerated, fluctuations in the data path—the sweep of the data points joined by the data line (Figure 1)—may be interpreted as being sudden and dramatic because of their extreme vertical movement. Contrarily, if the horizontal axis is exaggerated, fluctuations may be seen as slow, flat, and unimportant. It is not always possible or desirable to use square graphs. With care, a rectangle that is wider than high allows suitable plotting of a lengthy time series or one with many data points; one higher than wide may be better for a short time series or for rapidly changing data. The format should fit the data, not vice versa (Schmid 1983).

The rectangle may have quantitative scales on its vertical and horizontal axes (Figure 1); however, one of the axes may be labeled with a qualitative scale. Choice of scales should allow data to fill the data region as completely as possible, and the axes should end at or near the level of the outermost data points. In some circumstances it may be helpful to the reader if two different scales are used for the same data (e.g., a logarithmic scale on the left vertical axis and its arithmetic counterpart on the right). Comparisons of graphs placed side by side or one above the other are facilitated when scales on each graph are similar. If graphs that are being compared have different scales, that fact can be signaled by deploying rectangles along their top or right sides to show relative scale size for each panel (Figure 1; see also Cleveland 1985).

Tick marks (Figure 1) are placed on at least the left vertical and bottom horizontal axes, and on the other axes if that would help the reader. The range of the tick marks should encompass the range of the data. Tick marks (and associated lettering) should be kept to a minimum and should be placed to avoid interference with data points; lettering should be placed far enough away from the axes to avoid the merging of axis lines or tick marks and lettering after the figure is reduced for publication. There is no need to number every tick mark; an appropriate interval can be chosen to avoid clutter.

The more significant a graphic element is, the more prominent it should be visually. There is a "figure–ground relationship" such that the figure should stand out conspicuously from the ground. This is a matter of contrast or emphasis and can be facilitated by shape or form, size of symbols, weight or pattern of line, or position or orientation of elements within the graph. For example, in a line graph, the data paths (Figure 1) should be prominent in the data region and no less conspicuous than the coordinate axes. Multiple data points that coincide exactly can be represented by "spokes" or "sunflowers" (Figure 1). Plotted symbols that overlap without completely coinciding need to be visually distinguishable; it may sometimes be necessary to offset the symbols slightly, either vertically or horizontally (such details should be reported in the legend).

Data should fill the data region to keep unused "white space" to a minimum. On the other hand, irrelevant data should be discarded and superfluous
graphic elements (e.g., labels, notes, markers) should be kept from cluttering
the data region. A reference line (Figure 1) may help to highlight an important
value that must be seen across the entire graph, but should not interfere with the
data.

Subsets of a data set, or two or more overlapping data sets presented in the
same graph must be clearly distinguishable. This means that symbols and
shadings must be chosen and used carefully. MacGregor (1979) recommended
that a contrast of at least 30% in gray value be maintained between adjacent
tones of shading. Many patterns of hatches, cross-hatches, and grids can
produce moiré effects in which the design seems to vibrate distractingly. Even
equally spaced, unadorned bars and columns may vibrate. Tufte (1983) called
such vibrating patterns "chartjunk" and warned that chartjunk is proliferating
as computer graphics programs increase in use. However, his suggestion that
chartjunk be replaced with screens of varying density or shades of gray, or with
words, should be followed with care. The different densities or shades may be
difficult to discriminate (as noted earlier) and may not reproduce distinctly
when printed. Similarly, words may not be a satisfactory substitute for vibrating
designs if they clutter the chart. If gradations of shading are not distinct, or
if the data region is obscured by words, the data might better be shown in a
table.

Inasmuch as most drawings benefit from being reduced, they can be prepared
33% to 50% larger than final printed size. Shadings, line thickness, and size of
data points and inscriptions must be able to withstand the later reduction.
Similarly, there must be adequate spacing of lines or dots in hatched, shaded,
or dotted areas, and adequate spacing between letters and between lettering and
lines. Letters and numbers especially must be of a size to withstand reduction;
final printing size of lowercase letters should be at least 1.5–1.75 mm.
MacGregor (1979) provided the following rule governing weights of lines, based
on the minimum height of the lettering on the figure: measure the width of the
lower-case letter el (l) of the minimum size and make the X- and Y-axes one el
wide, the curve lines two or three els wide, and the tick marks 0.5 el wide (if
they will then withstand reduction). As a final check after a graph has been
prepared on drawing material of a standard page size (22.5 × 28 cm), use a
reducing photocopier to reduce the original to about 67%, then reduce the copy
by as much again (Cleveland 1985). The second reduced copy should be clear,
with open legible lettering and distinguishable symbols. This test might also be
useful when a rough draft in the proposed size has been made and a few letters,
numbers, and symbols of different sizes have been drawn in. The reduced copy
will show the suitability of line widths and weights, lettering size, symbol type,
etc., and allow for adjustment before a final copy is prepared.

Drafting paper or film, or illustration board, is a suitable drawing material (Zweifel
1988). Ideally, the same style and size of lettering should be used in all figures in a
manuscript. Although not all style manuals agree, words in phrases are probably
easier to read as initial uppercase and lowercase letters than as all capitals; capitals
can be reserved for scale labels that are only one word long. Press-on type and lines
are often unsightly or hard to apply in a straight line. When possible, a mechanical
drafting set or a lettering machine should be used—never a typewriter. If computer-

created graphics are used, a high-quality, continuous-line plotter or laser printer is essential to preserve figure quality.

Every figure should have a legend that is clearly expressed and succinct, yet comprehensive and informative. Such legends should stand alone, without need for reference to the text, if possible. This independence is attained if the writer answers the questions, "what? where? when? why?" in the legend and draws attention to important features. Error bars should be explained correctly—do they indicate standard deviations of the data, standard errors of a statistic, or a confidence interval for a statistic?

Material to be submitted to the printer should not be oversize, if possible; a maximum size of 22.5 × 28 cm permits easy handling and mailing, and reduces the possibility of damage to the graphic. Several guides are helpful: submit original copy if possible because photographs of original drawings may be blurred; identify all figures with your name and the figure number, and indicate the top of the figure if that is not obvious; give figures in printer's proof the same rigorous scrutiny that is applied to text and tables (the passage of time since the figure was last examined may allow mistakes to be seen clearly); and check to ensure that lines are sharp, that lettering is legible and not blurred or filled in, and that symbols are distinct. Correction or substitution of artwork at the proof stage is costly, and the author is liable to be billed for the cost.

Use of Figures in Fishery Journals

We wished to explore use of figures and tables by fisheries scientists in order to determine which figures are most common, the relative abundance of figures and tables per journal article, and the ratio of figures to tables. We examined issues of *Environmental Biology of Fishes*, *Fishery Bulletin*, *Journal of Fish Biology*, and *Transactions of the American Fisheries Society* published from 1983 or 1984 to 1987 or 1988. The numbers of articles scanned per journal ranged from 279 to 490, figures numbered from 1,111 to 1,911, and tables numbered from 727 to 1,374. On the average, articles had 2.9–5.0 figures and 2.2–3.6 tables among the four journals; the ratio of figures to tables ranged from 1.2:1 to 1.7:1.

For each journal, we classified each figure into 1 of 36 categories[2] and determined the number in each category. We determined the percentage of the total 5,797 figures that each category represented and grouped the categories into major divisions (Table 1). Line graphs were the commonest figure used, except the category of photographs and photomicrographs was slightly more common in the *Journal of Fish Biology* (that journal numbers nearly every photographic figure individually; the other journals group many photographs into composite

---

[2]Figure categories; asterisks denote categories in which we tabulated figures that included a time component—line graph: single*, multiple*; surface graph: single*, multiple*; logarithmic graph: semilog*, log–log*; ogive*; probit; vertical bar graph: simple*, grouped*, component (compound, 100%)*, bilateral*, moving range*; histogram (simple, grouped); horizontal bar graph: simple, grouped, component (compound, 100%), bilateral, moving range; box plot–range bars; kite figures; circle-pie graphs; scatter graph; schematic diagrams; map; photograph–photomicrograph; isopleth; three-dimensional plot; response surface; illustration; recordings and traces (spectra, kymograph record, etc.); matrices; miscellaneous (predominantly mixed figures of two or more types).

TABLE 1.—Relative and total proportions (%) of the more common types (≥5% of total) of figures published in four fishery journals in recent years.

Division and category of figure	Environmental Biology of Fishes (N = 1,111)	Journal of Fish Biology (N = 1,370)	Fishery Bulletin (N = 1,911)	Transactions of the American Fisheries Society (N = 1,405)	Total (N = 5,797)
Line graphs					
All	30	33	29	41	33
Single line	12	7	10	13	10
Multiple lines	16	24	18	25	21
Photographs and photomicrographs	13	35	9	7	15
Bar graphs					
All	16	12	11	13	13
Single and grouped	6	5	5	7	6
Histogram	8	6	5	5	6
Schematics and diagrams	14	7	9	10	10
Maps	6	5	13	8	9
Illustrations	8	1	11	2	6
Scatter graphs	7	3	4	5	5

figures). Bar graphs (vertical and horizontal) ranked second or third in abundance. Schematics or diagrams, maps, illustrations, and scatter graphs generally were less abundant.

For many categories, we determined whether or not time was a variable in the figures. Of 2,332 figures examined, 52% involved time. This corresponds with Tufte's (1983) claim that about 40% of scientific figures are relational (not time series).

We now describe the various types of figures used in fisheries papers and offer suggestions about their preparation. We illustrate several points with published material. This material sometimes has modest shortcomings but we avoided use of poor figures that would embarrass authors. Throughout the following, we refer to additional examples of figures published in the *Transactions of the American Fisheries Society* (TAFS) and elsewhere to supplement those we include here. We also report in parentheses the percent of the total number of figures represented by each division or category when we discuss that division or category.

Rectilinear Coordinate Line Graphs (33%)

Line graphs are perhaps the most familiar and flexible of the various types of graphs. They portray fluctuations or visualize trends in continuous variables. They show either cause-and-effect or sequential relationships (e.g., the behavior of a variable or a set of variables through time), display long series of data, allow for interpolation, and permit comparisons of multiple data sets.

The line graph is a two-dimensional figure based on the quadrant system, or Cartesian plane—see Cleveland (1985) for ideas on plotting three or more quantitative variables. The plane allows mathematical functions to be graphed, thus illustrating the behavior of the variables in the equation. The X-axis (abscissa) crosses the Y-axis (ordinate) at the zero point or origin. Because most line graphs employ only the first quadrant of the plane, the origin occurs at the lower left-hand corner.

A line graph has three basic components: a complete or partial rectangle, horizontal and vertical scales, and at least one data path. Scales and scale values run from left to right on the X-axis and from bottom to top on the Y-axis (Figure 1). Numerical values on the X- and Y-axes are usually multiples of 2 or 5, rather than a series like 3, 6, 9. . . . Large numbers on the axes should be reduced to 1, 2, or 3 digits, and the axes labeled with the appropriate multiple. However, the use of exponents in scale labels often reduces clarity. For example, does the Y-axis label "Estimated passage of coho salmon ($\times 10^3$)" mean that the data in the figure were multiplied by 1,000 before they were plotted? Or were they divided by 1,000 by the author so that the numbers on the axis would be reduced to a few digits for plotting, with the result that they should now be multiplied by 1,000 by the reader? To avoid such ambiguity, the quantity (here, thousands) should be spelled out. Zeros should always precede the decimal point in numbers less than 1.

For graphs in which two or more data groups are being compared, one can use multiple-purpose scales with care (Figure 2; additional example, TAFS 115:696), usually with no more than two scales and preferably with one on one axis and the other on the opposite axis (TAFS 112:429). The alternative to multiple-purpose scales is to break the graph into its component parts and graph the parts one above the other or side by side.

Scale labels should describe X- and Y-axes clearly and unambiguously (Figure 1). The label is placed to the left on the Y-axis and should be printed broadside so it reads sideways from bottom to top; it should never be printed with each letter right-side up and reading from top to bottom.

Graphics experts take two positions on the desirability of a zero line (i.e., a line that displays the origin). Schmid (1983) claimed that the "amputated" graph is deceptive because a line graph without a zero line is as misleading as a vertical bar graph with the lower parts of the vertical bars deleted; the result is a distortion of the comparative sizes of variables and an exaggeration of fluctuations in values. Both Tufte (1983) and Cleveland (1985) disagreed with this position; Cleveland showed examples in which the use of a zero line actually could degrade the resolution of data and create excessive white space. If a scale break is needed, he recommended that it be a full break with two parallel lines extending across the graph to the opposite axis. Numerical values on the two sides of the break should not be connected. We take the position that the origin should be shown, except for time scales. A break (double straight-line perpendicular to the axis, double slant-line, or modified Z-break; Figure 1) should be placed in the Y-axis slightly above the zero line or in the X-axis slightly to the right of the Y-axis when the data path would otherwise be so far from one or both axes that too much white space would be left. The position of the break should not interfere with the plotting of the rest of the data.

Placement of data values on the graph should be as accurate as possible, because readers may need to extract the values for further manipulation. Two criteria are suggested for placement of data values along the X-axis in a time series: (1) if the datum is an average figure or a total for a time period, place it in the middle of the time interval; (2) if the datum is a measurement made early or late in the time interval, place it on or near the left-hand or right-hand side of the time interval, respectively.

If two or more data paths appear (Figures 1, 2), they can be labeled or otherwise distinguished clearly and redundantly with distinctive line patterns and symbols. Do not try to include too many data paths on a line graph if they intersect or clump

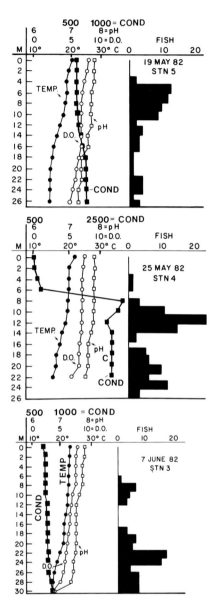

FIGURE 2.—Combined line graphs and horizontal bar graphs, with four scales (conductivity, pH, dissolved oxygen, and temperature) to describe the line graphs. Data for the environmental factors and fish abundance are plotted against increasing depth at 2-m intervals. From Matthews et al. (1985).

together. Data path labels should be simple and clearly understood; they should be printed horizontally and not along the curve of the path. Such labels are best placed next to the data paths if this can be done without cramping the graph. If that is not possible, group them in a special symbol key in an empty corner of the graph (Figure 1) where they will balance the design without clutter, or place the key just outside the figure's borders (preferably along the top of the figure), or define the data paths in the figure legend (a last resort). If you follow the last

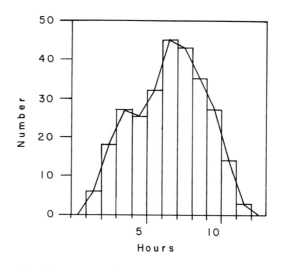

FIGURE 3.—Frequency polygon superimposed on a histogram.

suggestion, check the journal's instructions to authors or examine recent issues for symbols the printer can provide: odd or unusual symbols may not be available. Commonly, open and filled circles, triangles, squares, and diamonds are suitable; ×, +, and * are less suitable because they may not reproduce well. Cleveland (1985) noted that different types of circle fill can be satisfactory for encoding data. For example, an open circle with a thin circumference line, an open circle with a heavy circumference line, a closed circle, and a circle with a line across its middle allow for excellent visual discrimination. The only caution is to ensure that the circle with the heavy boundary will not fill in when reduced during printing.

If 95% confidence intervals are to be presented, they can be emphasized as shading along the data path. Such a display is visually effective (TAFS 110:757).

One type of line graph that shows the distribution of frequencies over a set of class limits is the *frequency polygon* (Figure 3). The first step in its production is to determine the appropriate size and number of class intervals. Too few intervals may make the classes so large that salient characteristics of the data are hidden; too many intervals may make the classes so small that some have few or no data. There are no precise rules about these matters, but Sokal and Rohlf (1981) recommended 12 to 20 size-classes as a sensible range to be used; the actual number selected will depend on the size of the data base and, to some extent, on one's familiarity with handling data.

Care is required when a frequency distribution is smoothed, because the smoothed curve is an idealized representation of the data. Smoothing eliminates irregularities caused by sampling error but it may not be possible to distinguish confidently between sampling error and true variation in the data. The smoothed curve can be superimposed over the original frequency distribution for comparison.

Surface graphs (0.4%) are line graphs that are filled in, the upper margin of each being a line graph (Figure 4). They can be visually misleading in that they make stronger statements than do line graphs. The filled-in area stands out clearly, and an illustrator can use variety and texture to provide unusual emphasis, especially

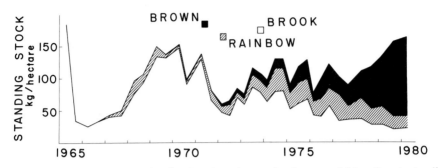

FIGURE 4.—Surface graph clearly showing change in three variables (trout standing stocks) over time. From Waters (1983).

for multiple surface graphs like the one in Figure 4. However, this emphasis may not always be appropriate in a scientific paper.

Data cannot be compared as readily with surface graphs as with line graphs. For example, surface graphs may be used to compare reproductive condition, compositions of trawl or plankton catches (TAFS 113:683; 115:719), or stomach contents (TAFS 114:522; 115:309) over time, the various components of each being represented by different shadings. However, only the lowest surface graph, representing one component, and the top one, representing the sum of all components, will be readily understood because they are built on the baseline. The other "layers" must be evaluated from a varying baseline (their zero line not being flat). Unless a surface graph has only a few (up to three?) layers (TAFS 115:809) or is as unambiguous as Figure 4, it might be best to use a multiple line graph in which all data paths have the X-axis as a zero base line. Or, a component or segmented vertical bar graph (discussed later) might be less confusing than a surface graph because the eye can focus on the results of one sampling period at a time. If a surface graph is to be used, the smoothest, least erratic curve is placed next to the baseline, the next least active curve upon that, and so on—unless there is a logical sequence to be displayed.

Three-dimensional or response-surface graphs (0.3%) are line graphs (Figure 5; TAFS 112:363). They must be prepared carefully so that the reader can estimate values from all three main axes. It is helpful to provide tick marks on all possible axes, and to indicate with dashed lines the positions of the axes that are hidden by the response surface. For authors with no access to an appropriate computer graphics program, commercially available triangular coordinate paper facilitates preparation of some of these complex graphs.

Logarithmic graphs (1%) provide an alternative to graphs based on arithmetic scales. When an arithmetic scale is used, equal distances represent equal amounts or increments of change. However, our ability to judge percentages and factors is poor. Thus, logarithmic scales should be used to show percentage change or effects of multiplicative factors; equal percents and multiplicative factors are represented by equal differences throughout the entire scale. A logarithmic scale often improves resolution of data that would be heavily skewed on an arithmetic scale. Cleveland (1985) discussed uses of logarithms in some detail and recommended the use of a $\log_2$ scale to provide such resolution. It is easier to deal mentally with powers of 2 than with fractional powers of 10; for this reason also,

Fall Collection

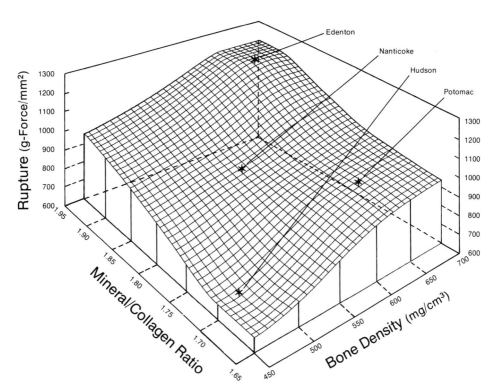

FIGURE 5.—Response surface graph of the relation between bone density, mineral:collagen ratio, and vertebral strength (rupture), for striped bass from four locations (means of the three characteristics at each location are shown by stars). From Mehrle et al. (1982).

the log$_2$ scale is useful even when the data range through no more than two powers of 10 (Cleveland and McGill 1985).

There are two kinds of logarithmic graphs. *Ratio or semilogarithmic graphs (0.6%)*, which have an arithmetic or equal-unit scale on the *X*-axis and one or more logarithmic cycles along the *Y*-axis (Figure 6), allow comparisons of proportional rates (or ratios) of change for two or more different curves, or for different segments of a single curve. The curves to be compared can include series with wide differences in absolute values. The *Y*-scale is known as the ratio scale because the rate of change of the data path is represented by the path's slope. The steeper the slope, the greater the rate of change; parallel curves show the same rate of change. If the slope of the data path is greater than, equal to, or less than 45°, the percentage change in the *Y*-variable is consistently greater than, equal to, or less than the percentage change in the X-variable (Figure 6). Finally, a semilogarithmic graph permits the comparison of series of data that are expressed in different units. If two percentage changes are to be compared simultaneously, *double-logarithmic or log-log graphs (0.4%)* can be used.

One can plot either the logs of a series of numbers on an arithmetic scale, or the actual numbers on a log scale. As suggested earlier, it may be helpful to have the logarithmic scale on the left vertical axis and the arithmetic scale on the right

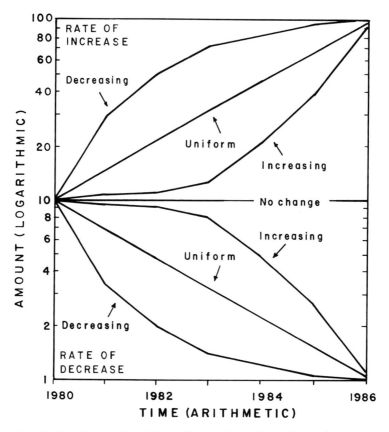

FIGURE 6.—Model of a semilogarithmic line graph to show shape of curves expressing rates of increase or decrease of the amount over time.

vertical axis—especially when the less common $\log_2$ or $\log_e$ scales are being used. There is no need for an axis break on the logarithmic scale because there is no zero base line; the interpretation of logarithmic curves does not depend on distance from a base line as it does for arithmetic graphs. Thus, the number of logarithmic cycles used depends on the values of the lowest and highest item of the plotted series, and partial cycles can be used.

Vertical Bar Graphs (13%) and Horizontal Bar Graphs (1%)

Conceptually simple, versatile, and readily constructed, bar graphs (Figures 7, 8) facilitate rapid transfer of information to the viewer. They allow comparison of numerical data over time or space, and of parts with the whole, and may be used to emphasize differences between elements. Percentage changes can be compared easily. Such graphs are seldom used for analytical purposes beyond simple comparisons; to interpolate between data points, for example, a line graph is needed.

A horizontal bar graph is most often used for size comparisons of descriptively labeled categories and thus may have only one scale; a vertical bar graph, or column graph, often compares one or more series of variables over time, and may have two scales (Schmid 1983). The vertical bar graph is like the line graph when

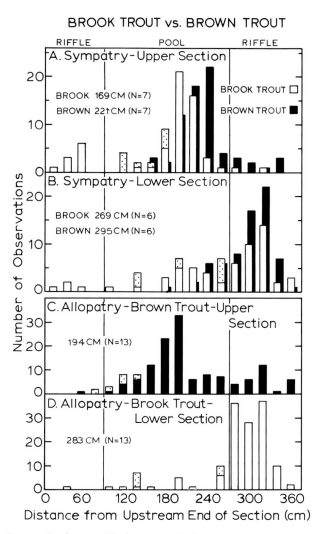

FIGURE 7.—Composite figure with simple vertical bar graphs in the lower two panels and grouped offset bars in the upper two panels (shading aids contrast). Data are for positions of trout in a laboratory stream; stippled portions of bars are the daily positions of dominant fish. From Fausch and White (1986).

it displays data in a single time series, but it gives greater contrast in the portrayal of data in two or three series (Schmid 1983). Horizontal bar graphs are appropriate for ranking variables, whereby the first or last ranking is at the top and the successive ranks are in descending or ascending order.

Data in bar graphs may be discontinuous or discrete; the bars then should not touch each other (Sokal and Rohlf 1981). Data may also be continuous (see histogram, described later) or they may be descriptive (e.g., names of fish) or qualitative (e.g., high stream discharge, low stream discharge). When a quantitative and a qualitative or descriptive component are graphed together, the quantitative value is usually measured along the X-axis of horizontal bar graphs and the Y-axis of vertical bar graphs. The following typology of bar graphs is modified after Schmid (1983):

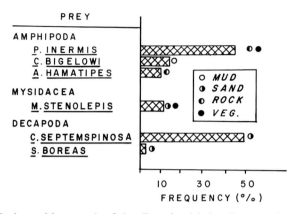

FIGURE 8.—Horizontal bar graph of the diet of red hake. Bars are balanced by textual information to left and a key to the substrate association of prey is inserted in otherwise empty space. Veg. = vegetation. Redrawn from Garman (1983).

The *simple bar graph (3%)* is a widely used form that displays a single series of variables or classes (Figure 7C, D; TAFS 115:571). Horizontal bar graphs may compare two or more coordinate items; comparisons are based on length and bars are ranked according to relative magnitude of each item (Figure 8) or some other factor such as water depth (Figure 2; TAFS 112:519).

Grouped or multiple-unit graphs (3%) allow comparison (Figure 7A, B) of two or three series of variables or classes within a series; the coding of more than three series by grouped bars is likely to result in confusion. The different series are differentiated by shading or cross-hatching. The grouped bars may be slightly offset (Figure 7A, B), may touch (TAFS 115:482), or may be separated by a narrow space.

Component or segment graphs (0.5%) can be used to display a series of component values in relation to the whole (Figure 9). Scale values are shown as

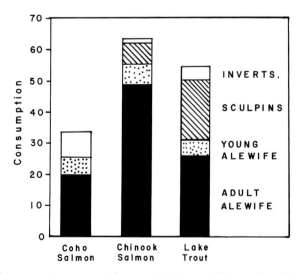

FIGURE 9.—Component or segment bar graph of prey of three salmonids, with consumption expressed as absolute numbers. Redrawn from Stewart et al. (1981).

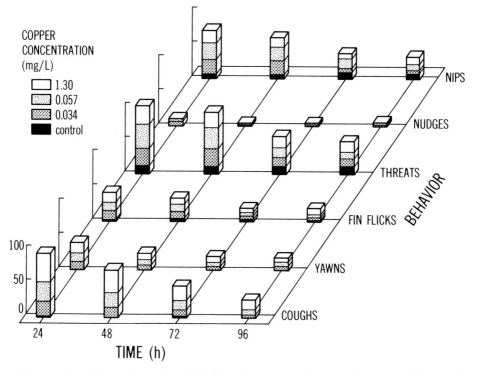

FIGURE 10.—Three-dimensional display of component bars to show mean frequencies of six behaviors of bluegills during 96-h exposures to three concentrations of copper or to control water. Vertical scale is mean number of occurrences (two replications) of a given behavior observed for a population for each time period. From Henry and Atchison (1986).

absolute numbers (compound bar graph, Figure 9), or as percentages of the whole (100% bar graph; horizontal, *Fishery Bulletin* 85:19; vertical, *Fishery Bulletin* 85:512). The largest component should be placed at the foot of each vertical bar or at the left end of each horizontal bar. Comparisons among components or segments become more difficult as the number of components increases. Thus, clearly differentiated shadings or line patterns are needed to help distinguish the components within bars (usually the darker shading is at the bottom of vertical bars or at the left end of horizontal bars). Or, lines can be used to join the tops of corresponding segments of neighboring bars to help emphasize trends and relationships. Component graphs can also be presented effectively in a three-dimensional form (Figure 10).

Paired or bilateral bar graphs (0.4%) allow for the simultaneous comparison of two quantities of several items or categories (Figure 11; TAFS 112:264). Deviation bar graphs compare positive and negative data or gains and losses (e.g., emigration and immigration of salmonids, TAFS 115:748). Vertical bars extend above or below the base or referent line and horizontal bars extend to the left or right of that line. One shortcoming of these graphs is that comparisons of the overall lengths of the elements (the sum of the lengths on both sides of the referent line) may not be easy. It helps to rank the elements by length.

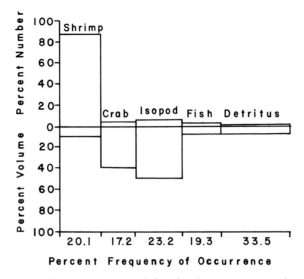

FIGURE 11.—Model of a paired bar graph for simultaneous comparison of volume and abundance of diet items in relation to frequency of occurrence (signaled by width of bars).

Moving-range graphs (<0.1%) show maximal and minimal data by use of vertical bars that "float" up and down along the X-axis (Figure 12) or horizontal bars that "swim" back and forth like fish (Figure 13; TAFS 112:324). There is a common scale but no zero line. The purpose of the graph is not to indicate how far each separate element deviates from a baseline, but rather how it compares to the other elements, perhaps especially to the one preceding it. Two sets of related information are shown by these graphs: lengths of the individual elements provide data on the increase or decrease of the total sum involved, and the comparative positions of elements on the scale yield data on performance over time.

The commonly used *connected bar graph or histogram (6%)* shows no space between contiguous bars (Figure 3), implying continuous data; the contiguous bars are interpreted as being part of a continuum. Thus, a histogram not only plots comparisons (as does a simple bar graph), but also displays relationships in terms of frequency on one axis and intervals on the other. Frequencies can be reported as percentages or absolute numbers. The criteria for deciding on size and number of class intervals are the same as for a frequency polygon (discussed earlier).

For each size interval in a histogram, it is assumed that there is a uniform distribution of units within each interval (represented by a bar width); this differs from the situation for the frequency polygon for which it is assumed that the number of units in each interval is concentrated at the interval's midpoint (Figure 3). The area of each class interval in a histogram is proportional to the frequency of that interval (if the class intervals are equally wide); this is not so for the frequency polygon. Thus, a histogram is preferable to a frequency polygon for graphic representation of a frequency distribution. On the other hand, there is a loss of information when a histogram is used, in that the width of a class interval is usually greater than the error or inaccuracy interval (due to measurement error and effects of rounding) of any data point included within that interval (Cleveland 1985). Two or more sets of data can usually be displayed more

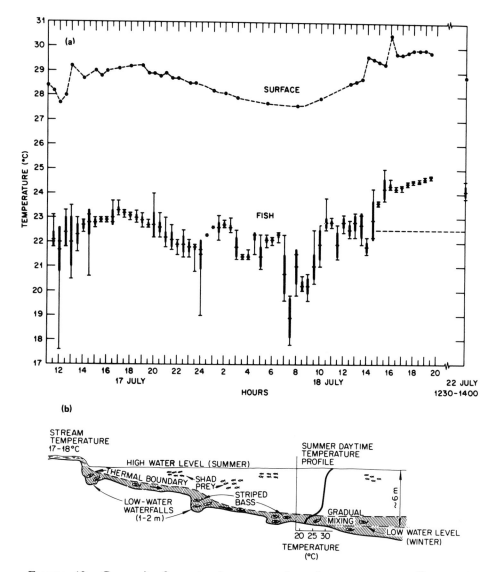

FIGURE 12.—Composite figure to show reservoir surface temperatures (line graph), temperatures selected by a tagged striped bass ("floating" range bars), and position of striped bass and their prey in relation to cooler deep water (schematic). From Coutant (1985).

satisfactorily by frequency polygons than by histograms because data may be obscured when histograms are superimposed. This problem may be preventable if angled histograms are used (Figure 14). One innovative histogram we encountered was composed of numerals, representing the scale age of single fish, that were arranged one above the other for each length class of two trout species (TAFS 112:472); the modal size classes were shown clearly to advance with size and age.

A break in the ordinate or vertical side of a histogram distorts the overall configuration of the distribution and should be avoided; it changes bar length and area—two key visual signals. It would be better to switch to a logarithmic scale to

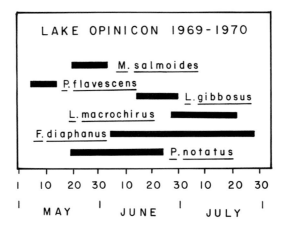

FIGURE 13.—"Swimming" range bars to show spawning periods of five fish species. Redrawn from Keast (1985).

allow comparison of all data. However, sometimes one bar is unusually long and dominant and would cause problems in drafting the rest of the bars. If there is no doubt as to its dominance, the bar can be broken neatly at a point beyond the next longest column (Figure 14).

Construction criteria for bar graphs.—For some data categories, the horizontal bar graph may be better if a ranking is involved. The ordering of a vertical bar graph in a time series is chronological, the earliest period being to the left of the graph.

Lengths of bars indicate data and should not be disproportionately long or short. If detail is unavoidably lost for some items represented by very short bars, a boxed insert may be added to an open area of the data region to provide an enlargement of the truncated data (Schmid 1983). Bars should not be too wide or too narrow and all should be of the same width in a single figure (except when bar width itself carries a message, as in Figure 11 or in an "area bar graph"). Usually, spacing between bars should not be more than about half their width (one-third is better). If bars touch in grouped displays, they may be allowed to overlap slightly to save space; however, one set of bars should then be longer than all, or nearly all, of the companion bars to provide a consistent and readable pattern or display. In relation to the body of a bar graph, at least two-thirds of the space should be allocated to the bars. In horizontal bar graphs, the longest bar should be placed on the bottom, if possible. If it must be on top, it is desirable to place something like a symbol key on or near the bottom of the graph for balance so the figure does not look as if it might topple (MacGregor 1979).

Some experts suggest that supplementary material should not be written inside a bar (Schmid 1983), but rather should be placed above or below a vertical bar (Figure 11) or at the left (or right) end of horizontal bars and close to the zero referent line (Figure 8). However, MacGregor (1979) warned that figures or words placed beyond the ends of bars sometimes create an illusion of greater height or length that hinders visual comparisons.

As noted for earlier graphs, shading should be done with care to avoid a cluttered appearance, optical illusions, or distractingly busy patterns. If all bars on a graph are rendered as unshaded outlines, or are equally shaded, all may be

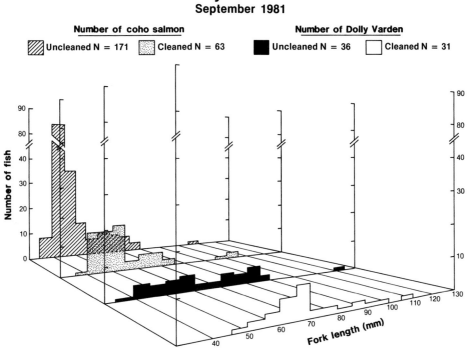

FIGURE 14.—Angled histograms arranged to allow comparisons, in a small amount of space, of length frequencies of two fish species in cleaned and uncleaned sections of an Alaskan creek. From Dolloff (1986).

perceived as being equal in weight. If one of a pair, or some of a group, are solid black, the black rectangle(s) attract the eye and give a disproportionate impression compared with that of unshaded neighbors.

Circle or Pie Graphs (1%)

Popular, simple in concept, and easy to understand, circle or pie graphs almost always deal with the relationship of parts to the whole, the parts being shown as percentages. However, these graphs have a low data density and can be misleading. Individuals drawing them may err when plotting along the circumference of the circle, and often make construction errors that exaggerate the areas of larger circles that are being compared with smaller ones. Areas of circles vary as the square of the half diameter, but illustrators may mistakenly base their drawings of the circles on the diameters, thus exaggerating the areas of the larger circles. Obviously, circles may be drawn correctly; however, because of the persistence of the drafting error just mentioned, one can never be sure if the comparisons are dependable unless the data are at hand, vitiating the use of the circle graph as a rapid means of communication. With the advent of computer graphics, such mechanical errors can be eliminated if software quality is high.

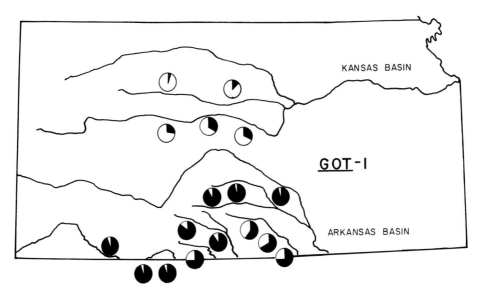

FIGURE 15.—Circle graphs with information on *GOT-1* allele frequencies of plains killifish within the state of Kansas (figure outline). Black areas within the circles represent the percentages of total genotypes surveyed that were heterozygous. Courtesy of K. Brown, Wichita State University.

Circle graphs are drawn so that the area of the circle or its segments can be used to compare items. However, it is difficult to compare such areas visually (see category "e" in the ranking of graphical perception tasks previously shown), especially if circles of different sizes are involved. In addition, it appears that readers may underestimate the comparative size of larger circles, even when the circles are drawn correctly. Thus, it seems best to use circles of the same size if one uses circle graphs at all. Only a few data can be presented (*Fishery Bulletin* 85:257), because too many segments (more than five or six) clutter the graph and make it unreadable. Circle graphs can be used satisfactorily in reports on two variables, such as the frequency of occurrence of two alleles (Figure 15; TAFS 112:8).

For single graphs, the division of the circle should begin at "12 o'clock" and the largest segment should be followed by decreasingly smaller segments. The smallest segments are best grouped near "9 o'clock," if possible, and no segment should be less than 5% (18°). If two or more circle graphs are used to track relative changes in quantities, the first can be drawn as suggested but it helps the reader if the remaining graphs maintain the order of segments of the first graph to allow for quick visual comparisons. Segments can be labeled or identified by a key. Labels are easier to read and to apply if limited to the horizontal plane. They can be put inside the graph if they remain large enough to be legible, or outside and adjacent to (or less desirably with arrows pointing to) the relevant segment. Circle graphs can be replaced by dot graphs (discussed later) or by the 100%-bar graph.

Scatter graphs (5%)

Conclusions can be drawn from the patterns shown by scatter graphs or scatter plots (Figure 1, panel 2), or these graphs may be used for preliminary data

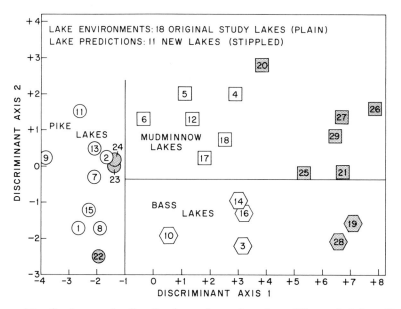

FIGURE 16.—Scatter graph of ordination values of northern Wisconsin lakes. Note the use of shading and shape to distinguish among groups of lakes. From Tonn et al. (1983).

explorations, or both. A variety of symbols or dot shapes may be used to differentiate data sets (Figure 16). For example, in a scatter graph of larval gulf menhaden length on age (*Fishery Bulletin* 86:84), numerals were used as plotting symbols, the numbers referring to the number of coincident data points.

If there is room, data dots may be identified with a horizontal verbal label; preferably all labels should be outside the data cloud to avoid clutter that would camouflage trends. Alternatively, the dots may be replaced by symbols large enough to hold a number (Figure 16) or letter that can be referred to in the figure legend (this procedure will affect accuracy in the plotting of data). Data of differing quantities may be represented by dots or symbols of different sizes. The result is more semiquantitative than quantitative because it is difficult to use area to estimate data size or value (Cleveland and McGill 1985).

Photographs and Photomicrographs (15%)

Photographic illustrations in journals are referred to as halftones, in contrast to the line illustrations (black lines on white paper) that we have discussed. Their reproduction involves the use of screens interposed between the author's photograph and the printer's camera. The screens reduce the images to more or less closely packed dots; it is the dots that are printed on paper. The better journals, especially those specializing in high-quality reproductions of scanning electron micrographs, use fine halftone screens.

Authors should submit only sharply printed, glossy photographs to be printed. It is best to place photographs on mounting boards and to protect their surfaces with a paper overlay. Do not write with pencil or pen on the back of a photograph; pencil impressions may mar the emulsion, and ink may bleed through the paper. Place instructions, your name, figure number, etc. on a press-on label that you then attach to the back of the photographs or mounting board. Indicate the top of

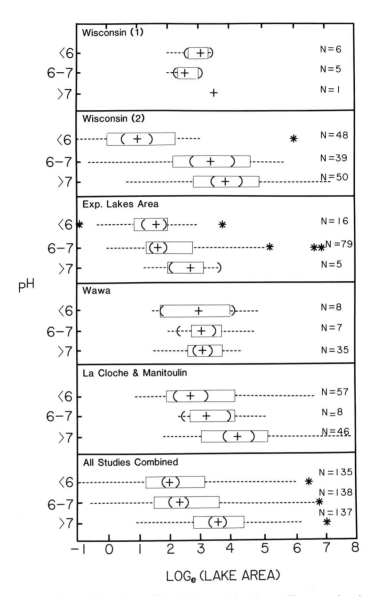

FIGURE 17.—Horizontal box plots of $\log_e$(lake area) by three pH categories, in which the median (+), its simultaneous 95% confidence interval (parentheses), the interquartile range (box), the range of adjacent values (dashed line), and outliers (*) are presented. From Rago and Wiener (1986).

the photograph. Keep fingerprints off the emulsion and do not use paper clips that might damage the face. To make plates from more than one photograph, cut square edges and butt the photographs together so that no white from the mounting board is visible; the printer will cut in white hairlines to enhance the appearance of the plate. If you need to crop a photograph, mark the mounting board at right-angles to the margin of the photograph. Always include any measurement scale within the photographs so that any reduction in printing equally affects the scale and the image. Place lettering on halftones so that white

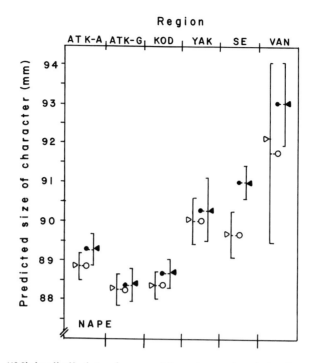

FIGURE 18.—"Minimalist" plots of geographic and sexual variation in measurements of nape length of 260-mm (standard length) Pacific ocean perch as represented by means of transformed (circles) and untransformed (triangles) data for males (solid symbols) and females (open symbols), and their 95% confidence intervals (brackets). Redrawn from Quast (1987).

lettering has a black background and black is on white; protect press-on lettering so that it cannot be knocked off accidentally (and examine proof meticulously to ensure that all lettering is intact). Final reduced size of lettering on photographs should be 2.5–3 mm. Allen (1977) provided greater detail on these matters. Finally, the production of photomicrographs requires familiarity with the microscope–camera system in use, and the appropriate instruction manuals must be studied carefully.

Miscellaneous

Because the production of *illustrations* (6%) involves a skill possessed by few scientists, the work is best left to a trained illustrator. However, Faber and Gadd (1983), Lindsey (1984), Zweifel (1988), and Hodges (1989) have provided helpful guides that enable a researcher to advise an illustrator not familiar with the specialized requirements for biological illustration. *Schematics and diagrams* (10%) may be more readily produced by scientists (Figure 12, TAFS 114:360) or computers. Various specialized templates may be used for drawings such as flowcharts.

Box plots and range bars (2%) can be helpful in graphing statistics economically (Figures 17, 18; TAFS 113:239; TAFS 116:833), and Hubbs and Hubbs (1953) improved on box diagrams for use in fisheries papers in systematics. Cleveland and McGill (1985) described another version of a box plot that may be useful to

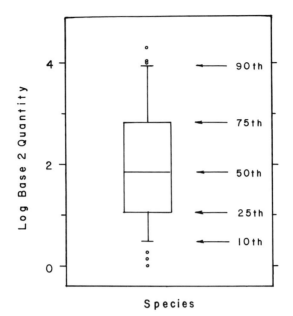

S p e c i e s

FIGURE 19.—Tukey box graph used to summarize the distribution of data and to show
the 10th to 90th percentiles of the distribution. All values in the data set above the 90th
and below the 10th percentile are plotted as circles. (See Cleveland (1985) for further
details.)

some fisheries scientists, especially when the presence or behavior of data outliers
is of interest (Figure 19). A mixed display of graphics can often provide many
complementary data in a compact space (Figure 12). *Kite diagrams* (0.1%) are
useful for displaying numerical data such as varying abundances of fish or
invertebrates over time (TAFS 113:221). *Isopleth diagrams* (1%) provide data
contours and may be shaded for emphasis (TAFS 114:485).

Graphic Advances

As we have noted, Cleveland (1985) and Cleveland and McGill (1985) have
provided helpful information on the relative accuracy of graphical perception
tasks. Cleveland (1985) discussed the use of spokes or sunflowers for multiple
data points (Figure 1), of rectangles adjacent to graphs to signal scale change
(Figure 1), of full scale breaks, and of $\log_2$ and $\log_e$ scales. One final innovation
involves the use of dot graphs to replace unsatisfactory displays such as circle
graphs, as well as bar graphs that display labeled data and depend on length as
a code. We use the data from Table 1 to illustrate dot graphs (note that the
presentation of the same data in a table and one or more figures is not usually
accepted by editors).

When data of Table 1 are presented in a component bar graph (Figure 20), the
comparisons of length that are involved may take more time than a simple perusal
of Table 1 would require. However, when the data are presented in a dot graph
(Figure 21), the comparisons involve graphical perception tasks at the top of the
accuracy scale (Cleveland and McGill 1985). These tasks are comparison of
position on a common aligned scale (the left two or the right two panels), and

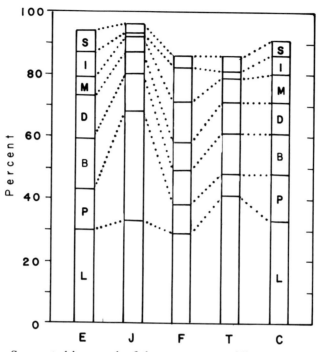

FIGURE 20.—Segmented bar graph of the percentages of line graphs (L), photographs and photomicrographs (P), bar graphs (B), diagrams (D), maps (M), illustrations (I), and scatter graphs or scatter plots (S) used in four fisheries journals in recent years. E = *Environmental Biology of Fishes*; J = *Journal of Fish Biology*; F = *Fishery Bulletin*; T = *Transactions of the American Fisheries Society*; C = all four journals combined.

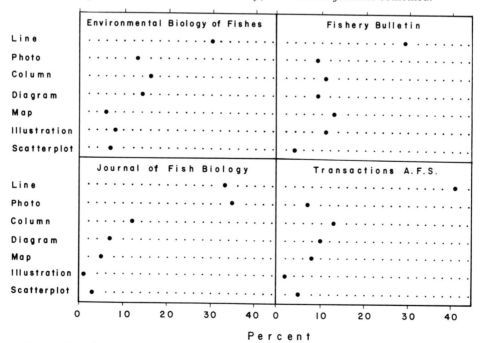

FIGURE 21.—Dot graph of the percentages of seven categories of figures in four fisheries journals. Data are the same as for Figure 20.

TABLE 2.—3-Way ANOVA.

Prey	Statistic	Fish species (1)	Treatment (2)	Turbidity (3)	1 × 2	1 × 3	2 × 3	1 × 2 × 3
1	F	40.6275	2.8162	.0034	11.1827	.5137	1.9384	.4662
	P	.001	NS	NS	.01	NS	NS	NS
2	F	.2139	1.6439	.1762	.0001	.9248	1.1624	.0476
	P	NS	NS	NS	NS	NS	NS	NS
3	F	11.1137	.0006	.3416	5.8062	.4167	.0102	.0096
	P	.001	NS	NS	.05	NS	NS	NS

comparison of position on common but unaligned scales (the upper two panels, the lower two panels, or the two in opposite corners of the figure). The shortcoming of such a dot graph is that it may require more room than the graph it replaces.

Although we have included light dotted lines from border to border (Figure 21), those between the data dot and the right axis can be omitted because the baseline is zero. In this instance, quantitative information is encoded by the relative positions of the data dots and the lengths of the dotted lines. If the baseline is other than zero, however, the light dots need to run from border to border because the baseline then has no particular meaning, the deviations from the baseline also have no particular meaning, and line length encodes meaningless data. Thus, the extension of the dotted line to the right border de-emphasizes the portions between the left border and the data dots, and the encoding is based on position and not position and length. This subject was treated more fully by Cleveland

TABLE 3.—Results of three-way analysis of variance to test the influence of fish species, treatment (i.e., species alone or together), and turbidity level on the frequency of different prey species eaten by young coho and pink salmon in feeding experiments (df =1). For presentation 1, F = calculated F-statistics; P = probability level; NS = not significant. For presentation 2, values are calculated F-statistics and asterisks denote $P < 0.05*$, $P < 0.01**$, or $P < 0.001***$.

Source of variation	Prey 1		Prey 2		Prey 3	
	F	P	F	P	F	P
Presentation 1						
1. Fish species	40.6	<0.001	0.2	NS	11.1	<0.001
2. Treatment	2.8	NS	1.6	NS	0	NS
3. Turbidity	0	NS	0.1	NS	0.3	NS
1 × 2	11.1	<0.01	0	NS	5.8	<0.05
1 × 3	0.5	NS	0.9	NS	0.4	NS
2 × 3	1.9	NS	1.1	NS	0	NS
1 × 2 × 3	0.4	NS	0	NS	0	NS
Presentation 2[a]						
1. Fish species	40.6***		0.2		11.1***	
2. Treatment	2.8		1.6		0	
3. Turbidity	0		0.1		0.3	
1 × 2	11.1**		0		5.8*	
1 × 3	0.5		0.9		0.4	
2 × 3	1.9		1.1		0	
1 × 2 × 3	0.4		0		0	

[a] The P-columns would be deleted from the table if presentation 2 were used.

(1985), who also showed how to present two-way, grouped, and multivalued dot graphs.

Tables

An advantage of graphs over tables is clearly revealed by four sets of bivariate data devised by Anscombe (1973). He developed a table in which each of the four sets of 11 X-values has a corresponding set of Y-values so that every data set has exactly the same values for 10 statistics (mean of Xs, mean of Ys, correlation coefficient, sum of squares, etc.). Careful, time-consuming scrutiny of the four sets of X–Y data in the table does reveal some odd patterns. However, the shapes of these patterns become immediately obvious when one sees the graphic plot of each data set. The data points of one X–Y set form a scatter graph with positive slope, the second a parabola, the third a straight line of positive slope plus an outlier far above the line, and the last a vertical line with an outlier far to the right. The viewer immediately grasps the graphic differences in the four plots that were masked in the tabular material. Why then use tables at all?

In most instances, tables provide exact numerical values, in contrast to graphics that portray the trend or shape of data, or the relation between experimental variables. However, because tables are expensive to compose and print, they should not be used unless repetitive data must be presented (Day 1983). If only a few data points are involved, they belong in the text in a few succinct sentences. This is also true for short word lists (e.g., numbers of fish harboring various parasites). If a sentence format results in ambiguity, a text table may be clearer.

The CBE Style Manual (CBE Style Manual Committee 1983) provides an excellent, detailed discussion on table construction that we will not repeat. However, there are some general principles to note; see also Ehrenberg (1977) and Day (1983).

Tables resemble graphics in that they should be constructed so patterns of data and exceptions to those patterns are obvious on quick inspection. This becomes easier if the table caption describes the general contents of the table. It is easier still when the reader is informed by a brief commentary in the text as to what the table illustrates. This text commentary should not be an account of each datum in the table; the essentials should be reported and the reader left to examine the details. Within the table, the message is obscured by inexplicable abbreviations, idiosyncratic format and spacing, inarticulate or excessive labels, and unnecessary digits.

For clarity, the caption, internal headings, and footnotes should complement each other, with the result that the table stands alone. A succinct caption (like a good figure legend) should contain all the information (and no more) needed to tell the reader what the table reports. Column and side (stub) headings identify the entries in the columns below the headings or in the rows to the right of the headings. Such headings must be brief but informative. Uncommon abbreviations can be explained in the caption or in a footnote, as can other material that would clutter and obscure the body of the table if included. Check the journal's instructions for acceptable footnote symbols. We will now follow a set of data as we improve its tabular presentation (Tables 2, 3).

It is easier to compare data down columns than across rows. Thus, Table 2 is more difficult to grasp than is Table 3. In addition, the caption of Table 2 is not helpful, the table shows too many decimal places, and there are no leading zeros. Table 3 has the data of Table 2 arranged in columns, the caption is informative, abbreviations are explained, and unnecessary digits have been deleted. Researchers are often loath to discard data or to round off to save space; after all, the data were not easy to amass. Nevertheless, there is no need to report many decimal places just because the computer does. Sokal and Rohlf (1981) dealt with the question of significant digits. Ehrenberg (1977) stated that data should be rounded only to two significant digits (final zeros do not matter because the eye is not distracted by them). Certainly the numbers 1,500, 2,100, and 8,900 are easier to manipulate in mental arithmetic comparisons than 1,526, 2,074, and 8,896. As in the lettering on figures, the leading zeros are required before decimal fractions less than 1.

Table 3 (presentation 1) is clearer than Table 2, but it is still cluttered. The data can be presented still more simply (Table 3, presentation 2). Again, data to be compared are in columns, the caption is informative, but unnecessary columns have been removed and pertinent symbols have been used as a code for statistical significance. Nevertheless, one must always consider whether any proposed table (and figure) is really needed. In our example, there are only four important items of data, i.e., those flagged as statistically significant. They can be reported in the text and the table can be abandoned.

Within a table, items should be grouped logically. A rule of thumb is to present control or reference data in a column to the left in the body of the table or in the top row. This establishes a reference to which other values can be compared. If a pattern established in a table is relevant to subsequent tables, it should be maintained. Similarly, the text that describes the table(s) should follow the same pattern.

Such consistency frees the reader's mind from the task of juggling among disordered remembrances by establishing the order we mentioned near the beginning of this work as a touchstone for graphic presentation. Indeed, the effort to order data clearly and simply in figures and tables forces an author to impose order in her or his own thinking, and this struggle facilitates preparation of the complete paper.

Acknowledgments

We thank B. Baldwin, P. Eschmeyer, R. Kendall, J. White, and two referees for their comments on earlier drafts of this manuscript. We are also indebted to G. Atchison, K. Brown, C. Coutant, A. Dolloff, K. Fausch, W. Matthews, P. Mehrle, P. Rago, W. Tonn, and T. Waters for use of originals of their figures. Finally, we appreciate the invitation from John Hunter to participate in such an important symposium. This is contribution 1980 HPEL from the Center for Environmental and Estuarine Studies.

References

Allen, A. 1977. Steps toward better scientific illustrations. Allen Press, Lawrence, Kansas
ANSC (American National Standards Committee Y15). 1979. Time-series charts. American Society of Mechanical Engineers, New York.
Anscombe, F. J. 1973. Graphs in statistical analysis. The American Statistician 27:17–21. (Not seen; cited in Tufte 1983.)
CBE Style Manual Committee. 1983. CBE style manual: a guide for authors, editors, and publishers in the biological sciences, 5th edition. Council of Biology Editors, Bethesda, Maryland.

Cleveland, W. S. 1984. Graphs in scientific publications. American Statistician 38:261–269.

Cleveland, W. S. 1985. The elements of graphing data. Wadsworth Advanced Book Program, Monterey, California.

Cleveland, W. S., and R. McGill. 1985. Graphical perception and graphical methods for analyzing scientific data. Science (Washington, D.C.) 229:828–833.

Cleveland, W. S., and R. McGill. 1987. Graphical perception: the visual decoding of quantitative information on graphical displays of data. Journal of the Royal Statistical Society A: General 150:192–229.

Coutant, C. C. 1985. Striped bass, temperature, and dissolved oxygen: a speculative hypothesis for environmental risk. Transactions of the American Fisheries Society 114:31–61.

Day, R. A. 1983. How to write and publish a scientific paper, 2nd edition. ISI Press, Philadelphia.

Dolloff, C. A. 1986. Effects of stream cleaning on juvenile coho salmon and Dolly Varden in southeast Alaska. Transactions of the American Fisheries Society 115:743–755.

Ehrenberg, A. S. C. 1977. Rudiments of numeracy. Journal of the Royal Statistical Society A: General 140:277–297.

Faber, D. J., and S. Gadd. 1983. Several drawing techniques to illustrate larval fishes. Transactions of the American Fisheries Society 112:349–353.

Fausch, K. D., and R. J. White. 1986. Competition among juveniles of coho salmon, brook trout, and brown trout in a laboratory stream, and implications for Great Lakes tributaries. Transactions of the American Fisheries Society 115:363–381.

Garman, G. C. 1983. Observations on juvenile red hake associated with sea scallops in Frenchman Bay, Maine. Transactions of the American Fisheries Society 112:212–215.

Gowers, E. 1977. The complete plain words, revised 2nd edition. Viking Penguin, New York.

Henry, M. G., and G. J. Atchison. 1986. Behavioral changes in social groups of bluegills exposed to copper. Transactions of the American Fisheries Society 115:590–595.

Hodges, E. R. S. 1989. Scientific illustration: a working relationship between the scientist and artist. BioScience 39:104–111.

Hubbs, C. L., and C. Hubbs. 1953. An improved graphical analysis and comparison of series of samples. Systematic Zoology 2:49–57.

Keast, A. 1985. The piscivore feeding guild of fishes in small freshwater ecosystems. Environmental Biology of Fishes 12:119–129.

Lindsey, C. C. 1984. Fish illustrations: how and why. Environmental Biology of Fishes 11:3–14.

MacGregor, A. J. 1979. Graphics simplified. University of Toronto Press, Toronto.

Matthews, W. J., L. G. Hill, and S. M. Schellhaass. 1985. Depth distribution of striped bass and other fish in Lake Texoma (Oklahoma–Texas) during summer stratification. Transactions of the American Fisheries Society 114:84–91.

Mehrle, P. M., T. A. Haines, S. Hamilton, J. L. Ludke, F. L. Mayer, and M. A. Ribick. 1982. Relationship between body contaminants and bone development in east-coast striped bass. Transactions of the American Fisheries Society 111:231–241.

Quast, J. C. 1987. Morphometric variation of Pacific ocean perch, Sebastes alutus, off western North America. U.S. National Marine Fisheries Service Fishery Bulletin 85:663–680.

Rago, P. J., and J. G. Wiener. 1986. Does pH affect species richness when lake area is considered? Transactions of the American Fisheries Society 115:438–447.

Schmid, C. F. 1983. Statistical graphics. Wiley-Interscience, New York.

Sokal, R. R., and F. J. Rohlf. 1981. Biometry, 2nd edition. Freeman, San Francisco.

Stewart, D. J., J. F. Kitchell, and L. B. Crowder. 1981. Forage fishes and their salmonid predators in Lake Michigan. Transactions of the American Fisheries Society 110:751–763.

Tonn, W. M., J. J. Magnuson, and A. M. Forbes. 1983. Community analysis in fishery management: an application from northern Wisconsin lakes. Transactions of the American Fisheries Society 112:368–377.

Tufte, E. R. 1983. The visual display of quantitative information. Graphics Press, Cheshire, Connecticut.

Waters, T. F. 1983. Replacement of brook trout by brown trout over 15 years in a Minnesota stream: production and abundance. Transactions of the American Fisheries Society 112:137–146.

Zweifel, F. W. 1988. A handbook of biological illustration, 2nd edition. University of Chicago Press, Chicago.

We Don't Care, Professor Einstein, the Instructions to the Authors Specifically Said *Double*-Spaced

ANDREW E. DIZON AND JANE E. ROSENBERG

National Marine Fisheries Service
Southwest Fisheries Center
Post Office Box 271
La Jolla, California 92038, USA

Abstract.—You can ensure that your relations with a scientific editor will be adversarial by following three basic rules. (1) Submit a manuscript that is inappropriate for the journal. (2) Prepare the manuscript in a style that deviates—but not consistently—from accepted standards of form, syntax, and spelling, and submit barely legible photocopies. (3) When critical reviews are received, treat them as an affront to your professional image and respond accordingly. On the other hand, relatively little effort by you will contribute much toward establishing and maintaining cordial relations with the scientific editor. Being objective about your manuscript and choosing the appropriate outlet, preparing your submission copy carefully, and dealing constructively with the review–revision cycle will more than pay back the time invested. This should be obvious, but an amazing number of authors fail to pay such attention, and thereby prejudice the acceptance of their manuscripts.

For most scientific journals, the editor wields absolute power. For instance, answering to no one, the scientific editor has the power to reject your manuscript without review ("Dear Author, Although your manuscript dealing with an application of the unified field theory to age and growth in fishes seems interesting, I am afraid I have recently received several manuscripts on exactly the same subject . . . "), the power to send your manuscript out to the most savage of reviewers ("Dear Dr. Legree, The author of the enclosed manuscript thinks it has merit . . . "), and the power to interpret the comments of a reviewer in the worst possible light ("The reviewer, in raising the remote possibility that sunspots affect . . . , forces me to request that you repeat . . . ").

Given the potential of such egregious application of power and the somewhat capricious nature of the review process, it behooves you to cultivate and maintain cordial relations with the scientific editor. Short of stuffing $100 bills into the submission package, how do you do that? I, from the vast perspective of almost 2 years as a scientific editor for the *Fishery Bulletin,* and my colleague will advise you how to maintain a productive association with the scientific editor and thus facilitate publication of your manuscripts. Our purpose is to examine the submission–review– revise–acceptance process, concentrating on what we see as your responsibilities as author. We deliberately avoid discussing the responsibilities of the scientific editor. After all, when you deal with dictators, it makes good sense to avoid hectoring them about their shortcomings. If you violate certain basic, commonsense rules, you will find yourself dealing with an unenthusiastic, hostile, or "imposed-on" editor; none of these personas is conducive to rapid publication of your manuscript.

Choosing the Appropriate Outlet

To ensure that the scientific editor will be enthusiastic about receiving your manuscript, you must choose the appropriate outlet. Along with an exponential increase in the overall number of journals (there are now at least 40,000 scientific journals around the world: Garfield 1988), the number of serials related to fishery topics has also increased. From 3 to 17 new fishery titles have been added per year, worldwide, since World War II (Maclean 1988). With such a large selection, it should be easy to find a journal eager to publish your work. Presumably you know the length and content of your manuscript. (And, of course, it is competent science.) The difficulty is judging objectively its value as a contribution. Is it a data report from the results of a baseline survey or does it describe a fundamental conceptual advance? That is, what is its contribution level? Objectively judging your own work helps you to select the outlet; it is hardly productive to spend time arranging the data report of your baseline survey in the format and style for submission to *Science*. Most authors do not make such gross errors in judgment; nevertheless many authors seem unsure of the value of their manuscripts. We spend a substantial amount of time "previewing" manuscripts in response to authors doubtful of their contribution's appropriateness for our journal. For all but the most novice of authors, this step should be unnecessary and tends to prejudice the editor. When authors ask an editor if a manuscript is appropriate for a journal, they usually seem to be saying, "This may not be the most interesting manuscript in the world, but would you consider it anyway?" Or, "I know this is in miserable shape, but do you think it is worth more effort on my part?" The best course is to dispassionately judge the quality of your own manuscript and carefully choose a journal that you know, from past articles, publishes manuscripts of a similar contribution level. Never let a scientific editor see a product of yours that is not up to the quality of most of the manuscripts that he or she sees.

Choosing the appropriate journal requires that you know something about it. Familiarity with its content should help you decide. We assume you have read it. Is your manuscript similar in scope and contribution level to those published in recent issues? The policy statements in the instructions to authors also help. Have you read them? We suspect that a considerable number of authors neglect to do so. *Fishery Bulletin* gets a surprising number of manuscripts that include original drawings even though the instructions clearly state that they should not be sent until requested.

These preliminary investigations allow you to decide if the level of contribution of your manuscript is sufficient for the journal. The scientific editor will then support its publication, if not enthusiastically, at least neutrally. These investigations also help you put your contribution in perspective with other articles previously published in the same journal. It is important that this is made clear to the scientific editor. "My article is a logical extension of the questions raised by Blank et al. published in 19XX."

Now that you have decided that the prospective journal publishes papers similar to your contribution, is the journal itself adequate for your purposes? At the outset, we suspected that there might be three areas of interest to authors regarding potential outlets for their manuscripts: (1) circulation, (2) number of manuscripts submitted and rejected per year, and (3) journal impact or influence. The relationships (or the unexpected lack of them) that we discovered are

TABLE 1.—List of journals from which information was obtained on page charges, circulation size, number of manuscripts submitted per year, and proportion of manuscripts rejected. (The scientific editors or editorial assistants were contacted by telephone or letter.) Journals marked with an asterisk impose page charges.

American Zoologist
Aquaculture
Biological Bulletin*
Bulletin of Marine Science
California Fish and Game*
Canadian Journal of Fisheries and Aquatic Sciences*
Canadian Journal of Zoology
Copeia*
Environmental Biology of Fishes
Fishery Bulletin
Journal of Crustacean Biology*
Journal of Fish Biology
Journal of Mammalogy*
Journal of Wildlife Management*
Journal of Experimental Marine Biology and Ecology
Journal of Shellfish Research*
Limnology and Oceanography
Marine Biology
Marine Mammal Science*
North American Journal of Fisheries Management*
Physiological Zoology
Progressive Fish-Culturist*
Science
Systematic Zoology*
Transactions of the American Fisheries Society*

intriguing, but the analyses have been left at a very superficial level. As a result, our interpretations are very speculative. Our goal was not so much to draw conclusions about the publishing process but rather to present information that an author might wish to investigate further and in greater depth for his or her own particular situation. We collected information by an informal telephone survey from the editors of various journals that American Fisheries Society scientists probably read and publish in (Table 1). The list is a casual selection of publications on our current journal library shelf. The editors or their editorial assistants were asked about page charges, individual and institutional circulation, number of manuscripts received per year, and the proportion of manuscripts rejected. We also consulted *Journal Citation Reports* (Institute for Scientific Information, Philadelphia, Pennsylvania) for data dealing with journal influence.

Page charges are like a regressive tax; their burden is greater on less well-funded programs. Consequently, investigators with poor funding may wish to avoid publishing in journals that impose page charges (Table 1). However, some editors of journals that impose page charges disclosed to us that an author can sometimes negotiate to have the charges reduced or eliminated.

Circulation may be important to an author wanting to do more than simply publish. Surveys show that scientists primarily read the journals to which they subscribe (G. Kean, Allen Press, Inc., personal communication), and neglect other journals in their institutional library. If this is true, authors seeking maximum exposure should publish in journals whose subscription fees are low and that consequently have developed large circulations to individual subscribers. In contrast are those journals that could be classified as "commercial," typified

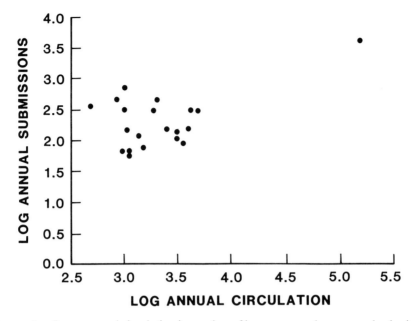

FIGURE 1.—Log_{10} annual circulation in number of issues versus log_{10} annual submissions in number of manuscripts for 20 scientific journals. *Science* is represented by the extreme point (total annual circulation = 150,000; total annual submissions = 4,000).

by *Aquaculture, Marine Biology,* and *Journal of Experimental Marine Biology and Ecology.* These are marketed to institutional subscribers, are very expensive (US$977, $1,683, and $1,300 in 1988 for the three journals just mentioned), and voluminous (24, 16, 30 issues per year). Circulation figures for these journals, and others like them, are confidential. Estimates of total circulation we obtained from a "knowledgeable source" were only 850 copies for *Aquaculture* and about 300 copies for *Environmental Biology of Fishes.* In contrast, *Science* has about 130,000 individual and 20,000 institutional subscribers. Specialty journals of societies, laboratories, and governments (at least in our survey) ranged from a high of 6,500 individual subscribers for *American Zoologist* to a low of under 50 for the *Canadian Journal of Zoology.* Most mature journals can assume an institutional circulation of 750 to 2,000 copies. Our journal, *Fishery Bulletin,* falls at the lower end; about 2,200 copies go to individuals and institutions combined.

We expected that scientists would want the maximum exposure possible for their articles and, therefore, would direct most of their manuscripts to journals with large circulations. However, for the journals in our survey (the ones for which we have information on submission and circulation), the number of annual submissions did not correlate with total circulation (Figure 1). Save for *Science,* with its large circulation and 4,000 unsolicited submissions each year, a large circulation does not seem to attract a large number of contributed papers. Seemingly, few scientists are concerned with circulation.

In contrast, rejection rate seems to be positively correlated with the number of annual submissions (Figure 2; $N = 20$, $P < 0.001$, $r^2 = 0.453$; Zar 1984). Again, *Science* provides an extreme value; of the 4,000 manuscripts received, editors can

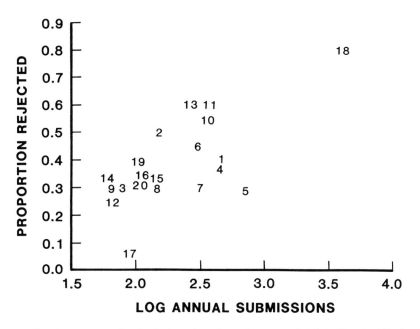

FIGURE 2.—Log_{10} annual submissions (number of manuscripts) to 20 scientific journals correlated with proportion of manuscripts rejected ($N = 20$, $P < 0.001$, $r^2 = 0.453$; Zar 1984). (1) *Aquaculture,* (2) *Biological Bulletin,* (3) *California Fish and Game,* (4) *Canadian Journal of Fisheries and Aquatic Sciences,* (5) *Canadian Journal of Zoology,* (6) *Copeia,* (7) *Environmental Biology of Fishes,* (8) *Fishery Bulletin,* (9) *Journal of Crustacean Biology,* (10) *Journal of Fish Biology,* (11) *Journal of Mammalogy,* (12) *Journal of Shellfish Research,* (13) *Limnology and Oceanography,* (14) *Marine Mammal Science,* (15) *North American Journal of Fisheries Management,* (16) *Physiological Zoology,* (17) *Progressive Fish-Culturist,* (18) *Science,* (19) *Systematic Zoology,* (20) *Transactions of the American Fisheries Society.*

reject 3,200 and still fill 52 issues per year. At the other extreme is *Progressive Fish-Culturist,* which rejects about 6 of its annual submissions of about 90 manuscripts. The *Canadian Journal of Zoology* deviates from the pattern somewhat in that it has 705 submissions per year and rejects only 29%.

Rejection rates for the sample of journals we used clustered around 30% (Figure 3). We do not know why. Perhaps scientists are capable of producing only two good manuscripts out of three, or, alternatively, every third scientist is incompetent. For *Fishery Bulletin,* we have no targets or guides on how many manuscripts we should reject, nor do we track or try to achieve any given figure. Nevertheless, we rejected 24% over the last 2 years, either immediately because they were inappropriate for the journal or after review. This percentage does not include "open" manuscripts, which require such extensive revision that they may never be resubmitted, or manuscripts that we requested be reduced to notes.

The influence of a journal can be estimated by various measures of the rate at which articles in that journal are cited. These analyses are provided by the *Journal Citation Reports.* One such measure is the "impact factor," a ratio of citations to citable items. It is the average frequency with which an article is cited during a particular year in other articles, both in the measured journal and in others. The measure for journal X for a given year, say 1984, is the number of 1984 citations

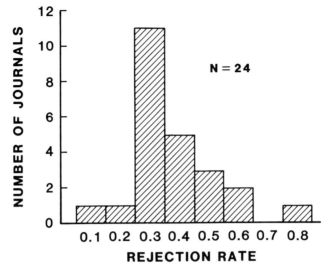

FIGURE 3.—Distribution of the proportion of manuscripts rejected by 24 of the 25 journals listed in Table 1.

everywhere to articles that had been published in journal X during the previous 2 years, 1983 and 1982, divided by the total number of citable articles published in journal X during those 2 years (Garfield 1972). The measure is biased by the size of the citing corpus and the choice of the 2-year window; this bias limits its utility for making comparisons between journals.

We examined the relation of impact factor to submission rate and, curiously, found that the two did not seem correlated (Figure 4). Once again *Science* is the exception, but journals such as *Limnology and Oceanography* and *Systematic Zoology,* whose subjectively perceived excellence is supported by high impact factors, are not distinguished by particularly high submission rates. Perhaps most scientists are unconcerned about the impact of their papers in terms of citability or, more likely, judge their contributions perceptively and avoid journals that would probably reject them.

Meeting Basic Standards

It is folly to increase the probability of a rejection by failing to meet the basic standards required of any scientific manuscript. By basic standards, we are not referring to scientific quality—that determination is somewhat more subjective. These are the standards, for the most part easily met, that dictate that a manuscript approximate the journal style, have easily readable text and prose, have legible and comprehensible figures and tables, and lack typographical errors. Nothing annoys a journal editor more than receiving a manuscript from an author who has obviously assumed that these details are unimportant.

The following admonitions may be reiterated by other authors in the present volume, but repetition will emphasize the point. Submitted manuscripts that meet the following criteria keep the scientific editor and the editorial staff reasonably happy and usually fare better than average in the reviewing process.

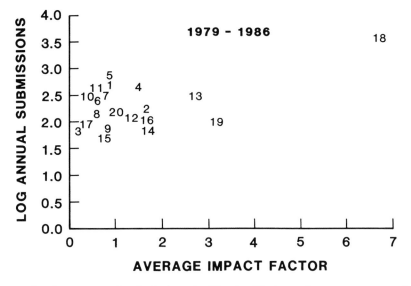

FIGURE 4.—Log_{10} annual submissions to 20 scientific journals versus average impact factor (number of citations per number of citable articles, from *Journal Citation Reports* 1979–1986). The following legend differs from that for Figure 2. (1) *Aquaculture,* (2) *Biological Bulletin,* (3) *California Fish and Game,* (4) *Canadian Journal of Fisheries and Aquatic Sciences,* (5) *Canadian Journal of Zoology,* (6) *Copeia,* (7) *Environmental Biology of Fishes,* (8) *Fishery Bulletin,* (9) *Journal of Crustacean Biology,* (10) *Journal of Mammalogy,* (11) *Journal of Wildlife Management,* (12) *Journal of Experimental Marine Biology and Ecology,* (13) *Limnology and Oceanography,* (14) *Marine Biology,* (15) *Marine Mammal Science,* (16) *Physiological Zoology,* (17) *Progressive Fish-Culturist,* (18) *Science,* (19) *Systematic Zoology,* (20) *Transactions of the American Fisheries Society.*

(1) The manuscript copy should approximate the style of the journal. It shows consideration for the copy editors and gives the scientific editor the impression that the manuscript was intended for his or her journal and was not rejected by some other journal. Approximating the specific style means that the copy editor will make fewer corrections on the copy. The more heavily a manuscript is marked, the more likely that errors will creep in during the typesetting process.

(2) The typography of the manuscript must be dark and legible and easily copied. Most manuscripts today are prepared on word processors; these should be output via a daisy wheel or laser printer. Do not submit a manuscript printed by a dot matrix printer; if you have no choice, use the emphasized or near-letter-quality mode with a new ribbon. Double-space all text, including the abstract, references, footnotes, and legends; they are edited too. Do not right-justify (i.e., do not create an even right margin); this adds extraneous spaces and makes the text harder to read. Do not hyphenate a word at the end of a line of typescript; ambiguous situations may occur. Indent each paragraph; if a new paragraph comes at a page break, it will be missed. Number each page; the frequency with which we get manuscripts without page numbers is surprising. Follow the guidelines of the American Mathematical Society for putting mathematics into print (Swanson 1979) and the CBE Style Manual (5th edition, 1983) for guidance.

(3) The manuscript must have readable prose (see Eschmeyer 1990 and Sinderman 1990, this volume). No more need be said.

(4) The manuscript must not contain unnecessary tables and figures (see Kennedy and Kennedy 1990, this volume). Ensure that the ones you include are comprehensible and well-designed. With the advent of inexpensive and easy-to-use computer graphics programs, more manuscripts include computer-generated figures and, perhaps as a result of the ease and low cost, include too many. Good biologists are not necessarily good graphic artists, and we believe that the quality of figures accompanying manuscripts has suffered as increasing numbers of biologists generate their own graphics with computers. Too many authors lose sight of the original goal of communicating information and get carried away with decorative graphical flourishes (fancy fonts, drop shadows, elaborate fill and grid patterns, shadow effects, exploding pie-wedges, etc.). Biologists producing their own figures should read "The Visual Display of Quantitative Information" by Tufte (1983), a remarkably readable and entertaining treatise on the historical evolution and proper design of statistical graphics.

(5) The manuscript must be proofread prior to submission. Take time with the preparation of the manuscript copy. The submission must be free of typographical errors and the style must be consistent throughout. Each manuscript copy must include all its tables and figures; although we at the *Fishery Bulletin* try to check each manuscript, incomplete copies occasionally are sent out to reviewers. We find it curious that a sizable minority of authors seems to feel that proofing copy is something that will be done by the journal editors, or that they will get to it after the review. They hope that reviewers will ignore these "trivial" matters and concentrate on the science. But sloppy manuscript preparation annoys the editors because it increases their work. It is also a red flag to reviewers; if the author is negligent with these details, reviewers often assume (and often correctly) that the carelessness carries over to the science. Reviewers are remarkably (and justifiably) harsh on authors of slipshod manuscripts.

In summary, the time spent carefully polishing a manuscript is trivial compared to the time spent collecting the data, analyzing them, and writing and editing the drafts. We think that a final polishing pays large dividends when a paper is reviewed and when the editorial decision about its publication is made.

Dealing with Reviewers' Comments

We assume that all scientists want the results of their research published, but we are not so naive as to think that their only motivation is a desire to communicate knowledge. Due to a variety of pressures, the act of publishing is itself a goal. We believe that it is important to recognize this as a modern reality; however, this need sometimes threatens the good judgment of authors about what constitutes merit in a contribution and makes them reluctant to accept negative criticism. ("I don't care if it's bad, I *have* to publish it.")

Ideally, authors should be as critical of their own manuscripts as anonymous reviewers are; however, this degree of self-criticism is unlikely. This unlikelihood and the pressures to publish makes an expert review system necessary. Thus, even if you are ill-served by a review, you still should recognize that the system is important, and treat the process with some degree of respect. It is not perfect, but better alternatives have not emerged.

Because most journals are multidisciplinary, their scientific editors depend heavily on the reviewers' recommendations. For the *Fishery Bulletin,* we attempt to get reviews from workers who are currently active in the subject field and who (we hope) are vitally interested in the subject matter. We trust that the solicited reviews are conscientious, informed, and thorough, because the verdict of the reviewers usually determines the final disposition of the manuscript. By and large, the system works. Our experience has been that most reviews are competent, and most authors find value in the comments of the reviewers.

Still, it is rarely pleasant to receive criticism, and criticism is sometimes applied with a heavy hand. From our observations, reviewers seem to develop a certain degree of hostility or righteous indignation toward manuscripts. This is understandable because reviewing manuscripts is an imposition with little reward. Most reviewers do not take on the persona of the wise teacher dispensing only constructive criticism. The perfectly normal, hostile emotions that the reviewer experiences fuel the search for error. The effectiveness of this hostility is borne out by the difference in quality between the reviews you the author solicit from your friends and colleagues (reviews that are often worthless or, if of value, too easy to ignore) and the ones supplied by the shadowy figures hiding behind the scientific editor. It is the anonymous reviews that reveal the weakness of the approach or the error in the methodology.

The fiendish joy that the reviewer experiences in finding flaws sometimes is apparent in the written comments that the author sees. The editor has considerable responsibility to maintain civility during the proceedings, but he or she is not always successful at winnowing out all potentially insulting comments. Some authors are insulted by the tone and hyperbole of such comments. This is somewhat puzzling; most authors have been reviewers, and many, if not all, have skewered an author or two themselves. In all situations, we suggest that the content of reviews, rather than the tone, be considered. Understanding what motivates the reviewer helps you ignore the hyperbole.

Besides resorting to snide hyperbole, reviewers sometimes are wrong and suggest poor revisions, and scientific editors sometimes pass these bad suggestions back to you. This gives you ample opportunity, in your resubmission covering letter, to indignantly pay back a career's worth of slights by all of the anonymous reviewers who have flayed your previous papers. Don't do it. Sometimes the scientific editor passes your sarcastic comments back to the reviewer—to whom you are not anonymous.

Handle the situation like this. Carefully evaluate all of comments of the reviewers and scientific editor. Handle those with merit, and look again at those without. In some instances, even wrong assertions or suggestions by the reviewers or scientific editor reveal a weakness in your argument or analysis. Reviewers are proxies for future readers. When they are confused, it is usually because of the author's bad writing. If, in a final dispassionate analysis, you conclude that the reviewer and scientific editor are simply wrong, be diplomatic even though this courtesy may seem one-sided. (After all, although the reviewer gets no real credit for a difficult job, you may end up with a published paper.) With your revised manuscript, you should write an accompanying letter indicating that you have carefully considered the reviewer's concerns and briefly noting how you handled each. For things that you are not willing to change (the objection is wrong, inappropriate, or stems from legitimate scientific differences), state why but keep the dialogue brief. We eventually published a manuscript dealing with fishery economics by an author whose rebuttal to a reviewer's comments was three times as long as the manuscript itself.

The scientific editor is occasionally caught in the middle of an apparently irreconcilable conflict between an author and an anonymous reviewer. In that situation, new reviewers are in order—some of whom might be suggested by you, the author. Additional reviews usually resolve the conflict; however, they can lead to contradictory advice, especially if the manuscript has gone through several review cycles. We have sometimes been embarrassed when the advice of new reviewers contradicts the advice that we have been dispensing on the basis of the original reviewers' concerns.

Sometimes the process goes totally awry. Our last warning to you is to avoid berating the scientific editor or journal about how poorly you were served by the process. In 2 years, we have had two irate authors who let us know exactly how they felt. Both swore they would never publish in the *Fishery Bulletin* again. One was mad because we rejected his manuscript. The other was mad because, although we accepted the manuscript, our reviewers were too slow at their jobs (actually, the delays were largely on the author's end). Revenge is sweet, but short-lived. Getting published is too important and too difficult for you to jeopardize future chances by eliminating outlets and alienating colleagues.

As we said in the beginning of this section, the peer-review process is not a perfect system—just a necessary one.

References

CBE Style Manual Committee. 1983. CBE style manual, 5th edition. Council of Biological Editors, Inc., Bethesda, Maryland.

Eschmeyer, P. H. 1990. Usage and style in fishery manuscripts. Pages 1–25 *in* J. Hunter, editor. Writing for fishery journals. American Fisheries Society, Bethesda, Maryland.

Garfield, E. 1972. Citation analysis as a tool in journal evaluation. Science (Washington, D.C.) 178:471–479.

Garfield, E. 1988. Too many journals? Nonsense! The Scientist 2(5):11. (Institute for Scientific Information, Philadelphia, Pennsylvania.)

Kennedy, V. S., and D. C. Kennedy. 1990. Graphic and tabular display of fishery data. Pages 33–64 *in* J. Hunter, editor. Writing for fishery journals. American Fisheries Society, Bethesda, Maryland.

Maclean, J. L. 1988. The growth of fisheries literature. Naga, the ICLARM Quarterly 11(1):3–4. (International Center for Living Aquatic Resources Management, Manila.)

Sindermann, C. J. 1990. Scientific writing as English prose. Pages 75–78 *in* J. Hunter, editor. Writing for fishery journals. American Fisheries Society, Bethesda, Maryland.

Swanson, E. 1979. Mathematics into type, revised edition. American Mathematical Society, Providence, Rhode Island.

Tufte, E. R. 1983. The visual display of quantitative information. Graphics Press, Cheshire, Connecticut.

Zar, J. H. 1984. Biostatistical analysis, 2nd edition. Prentice Hall, Englewood Cliffs, New Jersey.

Scientific Writing as English Prose

CARL J. SINDERMANN

National Marine Fisheries Service
Northeast Fisheries Center, Oxford Laboratory
Oxford, Maryland 21654, USA

Abstract.—Scientific papers, though forced into a standard format by prevailing logic and journal editors, can still be examples of excellent prose. To achieve such an objective, authors must invest time in good writing as well as in good science, because effective communication of findings is an important and inescapable professional responsibility. Good prose includes correct grammar, sensible paragraphing complete with topic sentences, adequate transitions from paragraph to paragraph, and introductions to all sections and subsections. Good scientific writing also includes effective use of abstracts and conclusions. Abstracts are best defined as "a few brief gem-like paragraphs that concentrate within themselves the essential qualities of larger works." Conclusions are oft-forgotten but critical components of scientific papers that state succinctly what the author considers to be the important contributions of the work. Then, as an overlay to all this correct writing, there is an elusive ingredient that transforms the document and gains the reader's attention. It may be the turn of a phrase, a spark of wit, a superb (but brief) summary of the literature, or a unique graphic presentation of data— something or anything that channels the paper away from the ordinary and toward the exceptional. Scientists should expect such writing from their colleagues, and should feel obliged to produce some of it in return.

The control they have over an important criterion of scientific worth— publication—sometimes entices journal editors down the dangerous pathways of arbitrary rule-making and unilateral interpretations of what is valuable in science. Most editors try to restrain any evil tendencies of this kind, and welcome manuscripts in which good science is reported lucidly. My contribution here focuses on lucidity—and on an expansion of that term to embrace its reluctant bed partners: clarity, effectiveness, elegant simplicity, and crisp, precise language. These to me are the essential ingredients of good prose writing.[1]

My advice is elementary—treat the preparation of a scientific manuscript as an exercise in writing English prose—but the simplicity of the advice lies in the statement of it and not in the achievement of it. The production of good English prose is complex and difficult, but it is an attainable goal for all of us, if we are willing to make the investment.

Standard scientific papers are often forced by current convention into a rigid but technically justifiable format of "introduction," "methods," "results," and "discussion." Some editors are more insistent than others about this, and some are absolutely dogmatic. Fortunately, other forms of scientific writing exist:

---

[1]Prose has been precisely defined as "a literary medium distinguished from poetry by its greater irregularity and variety of rhythm, its closer correspondence to the patterns of everyday speech, and its more detailed and factual definition of idea, object, or situation" (Websters Third New International Dictionary).

reviews, semipopular or popular articles, and books. These permit greater flexibility in structure and presentation. I am still convinced, though, that even within the most rigid of technical formats, good writing can be and should be achieved.

Because this paper is part of a symposium whose title seems to invite the preparation of lists and dicta, I here offer a few (actually six) of my own that should lead to an intimate physical union of scientific writing and good English prose.

• My first point is that *good prose does not flow easily from the brain to the pen or word processor*. Every sentence must be drafted carefully, then revised, prodded, expanded, slashed, kicked, worried over, and polished until it is a thing of impeccable beauty—and even then it should always be subject to further manipulation. Replace words, modify the structure, or eliminate the whole thing, but never, never settle for first drafts of anything. First drafts, from most of us (including editors), are usually untidy, grammatically terrible, and astonishingly imprecise. Revisions are absolutely crucial to good scientific writing.

• My second point concerns that often abused unit of structure, *the paragraph*. Paragraphs are delightful inventions designed to present cohesive material in succinct form, and to divide the text into logical subunits. But paragraphs are not easy to write because, in addition to their informational content, they are expected to have a topic sentence and to provide some transitions from what has gone before and to what will follow. It is my firm belief, supported by reading hundreds of manuscripts, that most fisheries scientists have little understanding of, respect for, or interest in the paragraph as a key structural element in their papers, except as a device to make the printed page esthetically more pleasing—which is probably the least of its values.

• A third point, relevant especially to longer papers, comprises *the suitable subdivision of the manuscript into sections*, the adequate *introduction* of each section, and the *transitions* from one section to another that provide essential integration. Sections are often necessary if the results of multiple experiments are being reported or if long-term observations are being summarized. The objective is clarity; it is aided by logical subdivisions of the manuscript and by use of the same sequence of presentation in methods, results, and discussion.

Introductions, in particular—both for a paper and for its sections—are too often slighted in scientific papers. Readers like to know where the narrative will take them; they like to be hand-carried gently into subject matter areas, not dumped abruptly into specifics. The effort required is small—a superb general introduction and then a short introductory paragraph for each major subdivision of the manuscript—but the rewards can be great in terms of the smoothness with which the writing flows. The advice here is again very simple; give readers a little background and preparation—give them a sense of time and place—then move into the substance of what you have to say, using logical subdivisions of that substance to keep the reader oriented.

This advice serves you, the author, as well as your future readers. Imagine yourself for a few seconds as an editor of a reasonably prestigious fishery journal, working late in your office at home. You drag out one final manuscript for the evening. The opening sentence of the so-called introduction begins: ''The research reported here has led to significant new information about pollution effects on fish stocks.'' Your question is ''Why?'' Why can't the author set the scene, talk to me briefly about the status of knowledge in the subject matter area, the research needs, and the proposed studies? Slavish adherence to pressure for brevity can be overdone, as it has been so often in oral presentations that begin with "First slide please." The science may be good, but the hour is late and you're in no mood to rewrite someone's paper, so back it goes for revision—regardless of the value of the research results.

• My fourth point addresses the uniquely personal use of language—*the writing style of authors*—that vast area beyond the mechanical structure of sentences and paragraphs and sections. Books have been written and innumerable college courses given on this elusive subject, but there is still much that can be learned by the average scientific practitioner. The topic of *writing style*—of *how* you write—is one that really invites advice and pontification from editors. I pass on few of the most relevant comments that I have heard or read recently.

(a) Use language that is "accessible."
(b) Avoid, at all costs, writing that is ponderous or pedantic or jargon-ridden.
(c) Read some of your paragraphs aloud and critically; if they sound impossibly (or even slightly) authoritarian, stuffy, wooden, apologetic, or patronizing, you have a problem (and your readers will too).
(d) Look at a scientific manuscript as a story to be told—succinctly and precisely, but interestingly as well.
(e) Attempt to convey a sense of enthusiasm for the research being reported; this is not easy to do within the confines of a technical report, but it is possible.
(f) Heed the admonition from E. B. White: "style takes its final shape more from attitudes of mind than from principles of composition."

• My fifth point concerns *the use of humor in scientific publications*. Generally—*don't*! Journal editors and reviewers will close ranks solidly against you, noting the word "dilettante" or worse epithets on their little 3×5 author reference cards or in their computer listing of authors. In a way, it is too bad that science should be portrayed in journals as such a deadly serious business, when actually there are many light, funny, even joyous, moments in the practice of science. Also, many scientists are innovative people with sharp minds, and clever humor is not repugnant to most of them (except, apparently, in technical journals). Fortunately, some semitechnical or nontechnical publications permit occasional flashes of well-conceived humor, and a few of the more general scientific magazines even publish entire articles with unusual perspectives on science and scientists. Additionally, there are a few publications, such as the *Journal of Irreproducible Results*, that welcome articles on the witty perimeter of science. According to author Daniel S. Greenberg, there is even talk of a *Journal of Rejected Manuscripts*—funded by the Foundation for Trivial Research—for papers (including those attempting a touch of humor) that did not survive normal editorial scrutiny by others.

Those of you who have dared to try a little humor will undoubtedly agree that it must be done infrequently and with consummate skill; otherwise it detracts from the paper and is vulnerable to immediate deletion by editors. Don't be afraid to try, though, and cite me freely if reviewers give you trouble or recommend rejection of your manuscript.

• My sixth and final point is that a section that should be a critical component of every manuscript—the *conclusions*—is slowly slipping into oblivion, apparently abetted by the apathy of most journal editors. Papers in almost all journals now seem to end with a vague section labeled "Discussion," which may go on interminably, page after page, but which only rarely contains a succinct statement about what the author concluded from the results of his or her labors.

I have done intensive library research, extending over almost an entire afternoon, on this revolting development. A statistically nonrobust examination of editorial practices and paper formats in zoological journals, beginning with the 1940s, has disclosed some startling evolutionary trends (and I reveal them here for the first time). "Conclusions" sections were common in the 1940s and 1950s, even though some were labeled "Summary." Gradually, though, especially in the 1960s and 1970s, the summaries were translocated to the front of the paper and relabeled "Abstract," leaving most published papers to dribble off with the often unsatisfying "Discussion" section.

I beg those of you who are producing technical papers: reverse this trend; reinsert a strong independent "Conclusions" section into every manuscript; argue with journal editors who would deny you this privilege; and encourage your friends to bring logic and substance back to the important closing message of scientific papers.

Use of the nondescript section heading "Discussion and Conclusions" should not be allowed as an escape hatch either; nothing less than a free and independent section boldly labeled "Conclusions" should be allowed. Try it, I think you'll like it.

So, with some modest attention to these six principal points—

(1) exhaustive polishing of each sentence;
(2) elicitation of the paragraph's potential beauty;
(3) logical subdivision of the text into sections, each with suitable introductions;
(4) distinctive writing style;
(5) ever-so-careful use of humor; and
(6) clear and separate statement of conclusions—

scientific papers can be as pleasant to read as short stories or poetry. They can be so exquisitely fashioned that reading them will bring tears to the eyes and a lump to the throat. We should expect excellent writing from all of our colleagues, and we should feel obliged to reciprocate with offerings of comparable quality.

The novelist Joyce Carol Oates concluded an award acceptance speech a few years ago with this great observation: "The use of language is all we have to pit against death and silence." I add that, as scientists, we should feel enjoined to use language just as effectively as we possibly can, as part of the basic commitment of professionals to the continuity and the pleasure of doing and communicating good science.

Statistical Aspects of Fishery Papers

In two papers that follow, the authors address statistical problems that they have perceived in published fishery research papers. The theme of the larger meeting that included our symposium was that fishery science is multidisciplinary. No fishery scientist can be expected to be expert in all the fields represented in the science. This is particularly true in regard to statistical and mathematical skills, which some have in abundance but others lack. Those who do not possess such skills should seek statistical and mathematical help in the preparation of their papers—but carefully, because miscommunication between a consultant and a consultee with disparate skills is frequent.

Editors are aware of and concerned about these problems. No editor wishes to publish a paper that is incomplete or in error. How the statistical quality of papers in fishery journals could be improved was the subject of some discussion during the symposium. Some of the suggestions made were (1) to provide more statistical training in fishery programs, (2) for editors to use as reviewers the younger, statistically trained members of the profession, and (3) for the journal to hire a statistical reviewer. The first of these suggestions will only be beneficial in the long term. To set the second of these in motion, editors need to work with academic institutions to develop a list of statistically trained persons. The third involves costs that must be considered. Alternatively, when the editor and referee are in doubt as to the quality of the analyses performed, the author could be requested to have the work certified by an expert in the appropriate quantitative field. This might require the author or the author's organization to pay a fee to the certifying expert. The certification, including the name of the person involved, would be included as a footnote. As a variant of this, the quantitative expert might be listed as a coauthor. Some recent experiences with medical publications, in which long lists of authors are common, show that some coauthors may not take seriously their responsibilities for the quality or accuracy of the paper. Thus, having a quantitative expert listed as a coauthor may not fully reassure the editor.

It is unlikely that a single solution will resolve all such problems. Each editor must determine what seems best, taking into account the needs and scientific level of the journal, the fraction of statistical or mathematical content of the papers submitted, and the budget. Editors probably should seek a candid evaluation of recently published papers to determine if this is a serious problem for their journal. In any case, editors need to seriously consider this problem and perhaps emphasize it in communications with the associate editors and reviewers.

JOHN HUNTER
Editor

Writing for Fishery Journals
© Copyright by the American Fisheries Society 1990

Statistical Problems in Fisheries

Douglas G. Chapman

*Center for Quantitative Science, University of Washington
HR-20, Seattle, Washington 98195, USA*

Abstract.—I consider a few broad classes of statistical problems that confront fisheries researchers and those who use and judge the fisheries literature. One of these classes concerns the choice of sample size. Sample size often is constrained by exogenous factors such as funding or logistic feasibility, but the effect of sample size on the reliability of the results can be expressed in terms of variances or confidence intervals. These statistics are straightforward when the research involves parameter estimation; when experimentation involves hypothesis testing, reliability can be expressed via probabilities of detecting evidence for alternative hypotheses. A second class of problems involves tests of the assumptions that underlie the design of studies and the analytical models by which results are interpreted. Examples are drawn from several common fisheries endeavors—treatment–control and mark–recapture experiments; age and catch-per-unit-of-effort analyses—to show that underlying assumptions rarely can be accepted without tests of their validity. A third class of problems concerns replication, which is essential both to test hypotheses and to obtain valid tests of statistical significance. Reviewers and editors should examine fisheries manuscripts with these problems in mind. Authors should demonstrate explicitly that they have addressed these problems and accommodated them.

Anyone who peruses a current fisheries journal such as the *Transactions of the American Fisheries Society* knows that certain mathematical skills are required. Most of the papers present some statistical analysis that usually has been carried out on a high-speed computer. Additionally, some fisheries research involves model building in which other branches of mathematics or computer modeling are used. A substantial fraction of fisheries research involves statistical analysis, and this fraction has been growing in size and sophistication during the past several decades.

It devolves on the editors and referees of fisheries publications to evaluate such analyses. It would be helpful if they had the answers to the following questions. When the experiments were planned, was there also a plan for the statistical analysis? Did the statistical analysis involve assumptions that could and should be tested? Were the assumptions spelled out and were the auxiliary experiments to test them carried out? Was the analysis appropriate to answer the questions that the experiments were designed to answer? Was the sample reasonable—that is, did it allow differences of interest to be detected or reliable estimates to be derived? Was the analysis carried out correctly? Are the statistical aspects described in sufficient detail to enable duplication and checking? Are the raw data given so that an alternative analysis can be carried out if one is felt to be appropriate?

Many of these questions frequently are not asked; if they were, the answers often would be "no." Some "no" answers have obvious reasons. For example, inclusion of raw data might make a paper unduly long and thus unacceptable for publication. The pressure of space may also explain brevity of the analysis or the explanation. The analyses usually are carried out correctly. Often, however, important assumptions are not specified and hence not specifically considered. Sampling procedure often is not clearly spelled out, making it very difficult for a reader to determine the population to which the results obtained from the analysis apply. Even more frequently, there is a less-than-adequate treatment of sample-size determination.

One might sample the research publications in fisheries and provide some examples for which these questions yield negative answers. This has been done in some fields and some of the resulting papers are worth studying by potential researchers, reviewers, and consulting statisticians. Here, I have chosen to consider a few broad classes of problems that are relevant to some of the aspects cited that pertain particularly to fisheries.

Sample Size

Many people are interested in sample size, judged from the frequency with which statistical consultants are asked about this aspect of experimentation and research. Sample size often is constrained by practical feasibility. For example, if a funding agency has identified a maximum level of expenditure for a study it supports, the researcher may feel there is little point in considering sample size beyond what is possible with the available funding. It still is useful, however, for the experimenter (and also the funding agency, although the latter may not ask) to know the level of detection expected from the study. In this respect, it is useful to distinguish two broad classes of studies or experiments: a class that involves only parameter estimation and one that involves hypothesis testing.

When the statistical analysis involves parameter estimation, it is customary to supply an estimated variance of each estimate, a confidence interval, or both; this guides reviewers and readers to the experiment's reliability. Because it is easy to specify the desired level of the variance or width of the confidence interval, these statistics also are frequently used to determine sample size before an experiment.

When the statistical analysis involves a test of a hypothesis, the situation is somewhat more difficult. The alternatives to the hypothesis tested usually form a range of values, and may involve "nuisance parameters" whose values must be known or estimated before a sample-size determination can be carried out. Such determinations involve more sophisticated analyses than do most tests, or they require special tables such as distributions of noncentral chi-square or noncentral t-values. In practice, however, the problem of sophistication usually can be avoided by use of approximations that involve only the normal distribution.

Although nuisance parameters will not be known in advance of an experiment, it may be possible to establish a range of values likely to encompass their true values. It is then easy to develop a table that shows the probability of rejecting the null hypothesis for various possible values of the alternative hypothesis.

For example, consider a situation involving linear regression for which the model is $Y = \alpha + \beta X + e$; as usual, the errors, e, are assumed to be normally distributed with a mean of zero and a variance of $s^2_{Y/X}$. The plan is to test the hypothesis $\beta = 0$ on the basis of observations of Y and seven X_i values; $X_i = 1$, 2, 3, 4, 5, 6, 7. The usual statistic is

TABLE 1.—Approximate power of the test of $\beta = 0$ for various values of β (slope of linear regression) and $s_{Y/X}$ (standard deviation of regression of Y on X).

	β		
$s_{Y/X}$	0.1	0.5	1
0.5	0.07	0.99	1
1	0.02[a]	0.53	0.99
2	[a]	0.11	0.53

[a] Power very small; approximation not satisfactory.

$$t = \frac{b}{s_b},$$

b being the estimate of the slope and s_b the standard deviation of b; that is,

$$s_b = \frac{s_{Y/X}}{\sqrt{\Sigma X_i^2 - n\overline{X}^2}}.$$

Here, $s_{Y/X}$ is the sample estimate of $\sigma_{Y/X}$, the population standard deviation, n is the number of X values, and $\overline{X}$ is mean X. For the Xs chosen, the denominator is $\sqrt{28}$. The hypothesis is rejected at the 5% level if the absolute value of t exceeds 2.57: $|t| > 2.57$. Now we can approximate the power of the test against some alternative β by calculating the normal probability

$$P\left[\frac{b - \beta}{s_b} > \left(2.57 - \frac{\beta}{s_b}\right)\right],$$

which is

$$1 - \Phi\left(2.57 - \frac{\beta\sqrt{28}}{s_{Y/X}}\right);$$

Φ is the probability that a standard normal deviate equals or exceeds the quantity in parentheses. This approximation neglects the left tail of the two-sided test.

At this point, one may feel stymied because there is no information on $s_{Y/X}$, the standard deviation from the regression, and uncertain about the alternative βs to be considered. However, a table of approximate rejection probabilities can be constructed for various values of β and $s_{Y/X}$ (Table 1). It is easy to extend such a table to other values of β and $s_{Y/X}$, which might be more appropriate for a different problem.

Alternatively, one may get a prescribed power, say 0.90, by writing

$$2.57 - \frac{\beta\sqrt{28}}{s_{Y/X}} = -1.28,$$

or

$$\frac{\beta}{s_{Y/X}} = 0.728.$$

Thus there is a very reasonable probability of detecting alternatives that equal or exceed $0.728s_{Y/X}$. Or, consider that the experiment has been carried out and the estimate of $s_{Y/X}$ is 2.38. Then it can be concluded that the approximate probability of rejecting the null hypothesis of 0.90, provided the slope β differs from zero by at least 1.73.

The preceding discussion deals with power—the probability of rejecting the null hypothesis—rather than with sample size. The sample size was determined by the choice of observations at the values $X = 1, 2, \ldots, 7$. If it is feasible to choose a larger set of X values, and a power of 0.90 is still sought, the sample-size problem reduces to the equation for N, the sample size:

$$\sqrt{N} = \frac{3.75 \, s_{Y/X}}{s_X}.$$

Here is needed not only the specific alternative β but reasonable values for s_X and $s_{Y/X}$. With these specified, N is determined. Although it is simple to discuss at least approximations to sample size for hypotheses relating to a comparison of test and control groups (or of two groups in general), other models that involve more complex statistical tests require more sophisticated treatment. However, most research now being carried out involves simple situations. Ideally, the experimenters will give consideration, insofar as possible, to sample size before the experiment or observations are carried out, and will also review the power of the test after it is completed. A reviewer who enters the scene later also should evaluate the power of the test, particularly if the authors have accepted a null hypothesis.

Testing Assumptions

Many tests of hypotheses involve assumptions. Increasingly, researchers are concerned about whether or not the assumptions are met; they do tests to check the assumptions, or they adopt alternative procedures that involve no (or less restrictive) assumptions. However, this is not the universal practice. In the following sections, I point out some classes of situations in which assumptions must and can be tested. All of these situations are important in fisheries work and have been identified in the literature; nevertheless, one can find recent published examples of failure to carry out proper procedures.

Comparative Experiments with Count Data

A broad class of experiments involves measuring the effect of a treatment, the measured variable being the number of "successes" in the treatment and control groups. My introduction to this problem came early as a statistical consultant.

In 1939, Harlan B. Holmes and Willis H. Rich of the U.S. Bureau of Fisheries began a series of experiments to measure the effects of Bonneville Dam (Columbia River) on mortality of downstream-migrating salmon smolts. They divided their available fish into two groups with distinctive fin clips, then released one group above the dam (experimental group) and the other below it (control group). The releases were made in various ways—some over the spillway, others into the turbine intake—and continued for several years. The aim was to compare, between experimental and control groups, the subsequent returns of adult fish to the river. These returns occurred over several years and in several fisheries.

In the late 1940s, Holmes attempted analysis of the data by applying simple chi-square tests, which seemed a reasonable approach. However, one of the paired releases showed a benefit (more returns) to the experimental group that was significant at the 5% level. It was hard to conceive how passing through a dam complex could enhance survival, and other explanations were sought. One explanation proposed was that variability was greater than that of a simple binomial model, which postulates that each fish in the test is an independent replication of an experiment with constant probability of success (or survival). One immediately sees how a binomial model could fail: the probability of survival may differ between individuals or survivals could be dependent events. Either possibility leads to an alternative model that cannot be handled by simple chi-square analysis. To carry out significance tests or to obtain confidence intervals, an independent estimation of the variance must be obtained. This is done most easily by having replicate experiments under identical conditions. The differences between such replicates form the basis of a variance estimate. Because Holmes and Rich had not attempted such replicates, there was no basis for such an estimate, and thus no basis for analyzing the results of the 15 experiments carried out over the 10 years of releases and recoveries. Consequently, the results were never published.

When Holmes and Rich began their work in 1939, the need for such replicates was appreciated by few statisticians and even fewer experimenters. Further, Holmes and Rich had very few marks at their disposal (only paired-fin clips), and it would have been difficult for them to effect much replication. Development of the magnetic coded wire tag has now eliminated this problem with such experiments on salmonids because this tag is uniquely distinguishable for the life of each animal and has an identical effect on each. Replication is now easy and should be a standard procedure. Nevertheless, replication is not always done. For example, de Libero (1986) statistically assessed the use of coded wire tags with chinook salmon and coho salmon in Washington State. Of several hundred experiments involving releases of these two species, only a few involved replicates. De Libero showed that, for the paired experiments that were replicated, the coefficient of variation ($g(X)$ = standard deviation/mean) was given by the function

$$g(X) = 124 + 34 \log_e[(1/X) + 0.0265];$$

X is the number of observed recoveries. It is easy to demonstrate from this function that a chi-square test of homogeneity would generally lead to rejection at the 5% level. In short, the binomial (or more generally, multinomial) model is inadequate for this situation. This inadequacy is to be expected in many similar situations, in which many environmental factors superimpose variability beyond that of random sampling.

Many experiments involve comparisons between a tested group and a control group, and it is appropriate to determine as a first step whether the binomial model is appropriate for their analysis. If there are several replicates of the control group and several replicates of the treatment group, a preliminary chi-square test can be carried out to test assumptions about the appropriate model; then an appropriate test can be made. However, such replicate tests often are not carried out.

Conversely, I can point to papers in the *Transactions* by authors who performed experiments with replication but pooled the results of the several replicates. Perhaps the replicates were homogeneous and a test would have validated the binomial model and justified the pooling. However, because the test was not performed and the separate results were not shown, the reader has no way of evaluating the matter.

I introduced this problem by referring to experiments at Bonneville Dam, where the measured variable was survival to maturity as indicated by recapture of fish in a fishery or spawning area. In such experiments, it is necessary to sample extensively in time and usually also in space, and often to wait several years for the returns of mature fish. The importance of this problem in studies of salmonids on the Columbia River and some other systems in western North America has led to another approach to this problem. When comparisons are made of young salmonids released above and below upstream dams, survival is estimated by sampling at one or more dams farther down the system.

I do not further discuss these experiments but note only that the analysis of such experiments was covered in depth by Burnham et al. (1987). They emphasized the importance of models as a basis for the analysis of such experiments, and also emphasized the importance of replication. They devoted a full chapter to the topic and, among other things, noted a procedure for estimating the variance-inflator factor when replication was not built into an experiment. This procedure requires some supplemental information that might be obtained, for example, from subsamples in the recapture process. Of course, this is much inferior to designing replication into the experiment. Burnham et al. also devoted a chapter to ''Planning Experiments,'' which should be carefully read by anyone involved in experiments of this type.

Mark–Recapture Experiments

Over the past several decades, mark–recapture experiments have been widely used, and the analysis of such experiments has become increasingly sophisticated. This sophistication has been developed in part to extract more information from the experiments, but in larger part to test assumptions that are basic to the procedures or to obtain results that do not depend on the assumptions. The assumptions vary with the model but at least three are important for most analyses: (1) no mortality results from the tagging process; (2) no marks are lost before fish are recovered; and (3) all fish recaptured with a tag are reported. Auxiliary experiments provide tests of these assumptions and estimates of correction factors if the assumptions are violated (Seber 1982). Because failure of the assumptions means that the estimates obtained are biased, it should be standard practice to carry out such auxiliary experiments as part of the protocol of a mark–recapture experiment. Such auxiliary experiments include holding newly marked fish in live tanks to assess short-term tag loss, grading tagged fish at release to estimated vitality, double-tagging fish to determine long-term tag loss, offering rewards on some tags to gauge reporting frequency, and subsampling recaptured fish in such a way as to ensure accurate information on the absence or presence of a tag. These auxiliary experiments have problems of their own, but they provide information that is essential if the main experiment is to be meaningful.

Despite this obvious need, many results of mark–recapture experiments are reported with no attempt to resolve the problems identified. Occasionally, an author notes these problems, but nevertheless provides estimates without evaluating their quality. This is true not only for Petersen-type estimates but also for the important class of Jolley–Seber estimates, as Begon (1983) has demonstrated.

Age Data and Age Analyses

Aging of fish has a long history and has been extensively discussed. Age data are used in a variety of ways—to construct growth-at-age tables, to develop growth-at-age models that may be used in yield calculations, to calculate mortality or survival rates, and occasionally to estimate recruitment rates or numbers. Age data are also required for the widely used methods of virtual population or cohort analysis.

Because age data are so widely used, the reader of a fisheries paper with age data may unthinkingly assume that the data are correct as presented. For some species and in some situations, the ages are undoubtedly read correctly, but recent studies have shown that this is not always true. This is emphasized by a symposium volume (Summerfelt and Hall 1987) that includes several papers on variability, error, bias, and validation. Further, for some species, it is clear that both random error and bias increase with the age of the animal. Some of the problems of aging marine species have been evaluated by Lai (1985) and Lai and Gunderson (1987).

What are the implications of errors in age determination? Barlow (1984) demonstrated that estimates of mortality or survival can be seriously biased by erroneous ages. Lai and Gunderson (1987) also studied the effects of aging errors on estimates of growth, mortality, and yield per recruit for walleye pollock. Some parameter estimates are not seriously affected, they found, but such others as optimum fishing mortality and optimum age at first capture could be substantially affected. The effect of errors has not been evaluated for other uses of age data.

In most of the relevant literature, the validity of age determinations is discussed slightly or not at all. It is clear that greater effort should be applied to assure that age readings are correct and, particularly when large-scale age-reading operations are undertaken, that quality control procedures are used to minimize routine errors and to eliminate bias.

Even when age determinations are correct, however, the analysis based upon them may be misleading. Several methods to analyze curves of catch at age (e.g., Chapman and Robson 1961; Robson and Chapman 1962) have been developed under the assumptions that the population is stationary and that the catch is a random sample of the population. The assumption of randomness is usually restricted to situations in which there is a sigmoid selection curve such that fish above a certain size (hence, age) are fully available to the catching gear. Tests are available to check this assumption, though if the selection curve does not follow this simple model, the methods used to estimate mortality or other parameters are invalid. The assumption of constant recruitment is a more serious one and cannot be tested without additional information. If variations in recruitment are purely random, mortality estimates will be unbiased—but variances of the estimates may be underestimated if the model used is incorrect. If there are both random variations and trends in recruitment, mortality estimates may be invalid. If there is trend alone, trend and survival will be confounded and the resulting geometric model will look superficially satisfactory but yield an estimate of survival rate divided by trend rate rather than of survival rate alone. A full discussion of these problems—the confounding of mortalities, selection probabilities, and recruitment—was given by de la Mare (1988). Thus, a referee or an editor who receives a paper that includes an analysis of age data should consider not only the accuracy of the age readings, but also the extent to which assumptions underlying the analysis can be justified and what their failure implies for the results.

Catch per Unit of Effort

The use of catch per unit of effort (CPUE) in population dynamics studies is both widespread and long standing. It is usually based on the simple equation $C = qfN$; catch (C) is proportional to the product of effort (f) and average population size (N) over the catching period. Here, q is the constant of proportionality or the catchability coefficient. Users of the equation have been appropriately concerned about proper measurements of effort. However, even if effort is measured correctly, it is not at all clear that CPUE is proportional to population size at all population sizes and levels of effort.

Early applications of CPUE were made to trawl fisheries, where it seemed reasonable to assume (in the absence of gear saturation) that catch in a unit area swept by the trawl was proportional to the density and hence local abundance of fish. Problems with this assumption were demonstrated in a 1963 symposium held by the International Council for the Exploration of the Sea (ICES; Gulland 1964). That symposium included the seminal paper by Paloheimo and Dickie (1964), one of the early theoretical studies on the subject; others have followed, including MacCall's (1976) study of a sardine fishery and Clark and Mangel's (1979) analysis of tuna purse seine fisheries. A wide range of problems with the use of CPUE for pelagic fisheries was exposed in a 1979 ICES symposium (Saville 1980). For example, Ulltang (1980) reported no significant correlations between stock sizes from virtual population analyses and those derived from CPUE of purse seine fisheries. Later, Cooke (1985) modeled CPUE as a measure of whale abundance, and pointed out that purse-seine fishing for schooling pelagic fish is "similar in structure to whaling where the school can be likened to an individual whale." These theoretical studies demonstrate the complexity of the problem; among other things, catch depends on the behavior of both the fish and the fishermen. Similar conclusions were drawn for sport fisheries by Hackney and Linkous (1978) and by Peterman and Steer (1981).

A list of problems that arise in trying to relate abundance or density to some measure of CPUE would be long and perhaps not too helpful. It appears that users of CPUE must look beyond the correct measurement of effort and attempt to determine the true relationship between CPUE and population density by theoretical or empirical studies (or both). It is tempting to reason by analogy from one fishery to another—to assume that if a relationship between CPUE and fish density holds in one fishery, a similar relationship will hold in a different fishery. Such extrapolation is unlikely to be valid if there are differences in the behavior of the fish or the fishermen, or there are biological or physical differences.

Replication

One other topic merits consideration both because it is a serious problem in fisheries research and because it has been widely discussed in the ecological literature. I have referred to the need for replication when experiments that compare a treatment and a control are based on counts of "successes" and the replication provides a test of the binomial model's validity. If the binomial model is not valid, replication is essential to determine the statistical significance of the compared counts or confidence intervals for the response effect.

There is a broader class of experiments for which replication is basic to the experimental design; this essential aspect of experimental design goes back to R. A. Fisher (1951). It is the essence of what is needed to obtain a valid test of significance.

Hurlbert (1984) devoted a substantial paper to the topic. He defined "pseudo-replication" as the use of inferential statistics to test for treatment effects with data from experiments in which either treatments were not replicated (though samples may have been) or the replicates were not statistically independent. Pseudoreplication occurred in 27% of the 176 experiments published between 1960 and 1984 that Hurlbert reviewed.

Ideally, experimenters should consider all the variables likely to influence a treatment effect when they test for the significance of the effect or estimate its level. This can be done in two ways: a systematic design can be imposed so that other variables can be accounted for, or the experiment can randomize with respect to the extraneous variables.

For example, suppose an investigator is interested in the effect that introduction of a potential forage species will have on the growth rate of a game fish in a lake. Many factors affect this growth rate: other fish species in the lake (competitors, predators, other prey), chemical and physical properties of the lake, and so on. The experimenter may select the lake where the experiment is to be carried out or the selection may be dictated by extraneous factors. Growth rates are determined before and after the forage species is introduced and compared by some appropriate statistical test. If there is a significant difference between observations before and after the introduction, it cannot be inferred that this difference is due to the treatment, though it may be tempting to do so. The difference may be due to changes in other conditions in the lake or in the climatic regime. Hurlbert (1984) would call this an experiment involving pseudoreplication. The problem can be handled in several ways. For example, one might replicate over time; if the treatment is not reversible, however, the untreated or control situation must always precede the treatment measurement, which introduces a possible bias. Alternatively, one might replicate over space. Several areas are selected: some receive the treatment, the others remain as controls. Ideally, the treatment and control areas are selected randomly.[1]

---

[1]Random selection of treatment areas usually is impossible in applied ecology; one cannot decide the location of a dam or a power plant on a statistical basis, for example. In such a situation, it is easy to understand Eberhardt's (1978) statement that "field experiments in ecology [usually] either have no replication or have so few replicates as to have very little sensitivity."

TABLE 2.—Experimental design: potential sources of confusion in an experiment and means for minimizing their effects (after Hurlbert 1984).

Source of confusion	Features of an experimental design that reduce or eliminate confusion
Temporal change	Control treatments
Procedure effects	Control treatments
Experimenter bias	Randomized assignment of experimental units to treatments Randomization in conduct of other procedures "Blind" procedures[a]
Experimenter-generated variability (random error)	Replication of treatments
Initial or inherent variability among experimental units	Replication of treatments; interspersion of treatments; concomitant observations
Nondemonic intrusion[b]	Replication of treatments; interspersion of treatments
Demonic intrusion	External sacrifices, etc.

[a]Usually employed only where measurement involves a large subjective element.
[b]Nondemonic intrusion is defined as the impingement of chance events on an experiment in progress.

In an experiment like this, it does not help to take subsamples either in the control or the treatment situation. Subsamples are useful to measure another source of variability within area or time, but the major comparison between the treatment and the control is still confounded by time and area and perhaps other factors.

Hurlbert (1984) summarized potential sources of confusion in an experiment and the means to minimize their effects (Table 2). He suggested that statisticians who teach give much greater emphasis to experimental design, and that statisticians who consult be more hard-nosed and suspicious. I concur with both suggestions. Consulting statisticians and experimenters fail to communicate; such failure can occur anywhere in statistics, but is particularly likely to occur with respect to statistical design when all details of the experiment are not made fully clear.

Journal editors and referees often are in a quandary over statistical design. Should they go back to the author and seek clarification of the experimental design so they can determine whether it was proper for the hypothesis proposed and the conclusion drawn? If yes, they risk lengthening the paper and perhaps bringing in technical details of specialized and only limited interest. If no, they risk publication of an inadequate paper. Later, they may encounter another paper that purports to address the same hypothesis but shows contradictory results. Hurlbert (1984) suggested that the problem should be solved by "a barrage of clearly explained rejection notices." He felt that statistical sophistication is not the main problem; rather, most experimental designs in field ecology are simple and "when errors are made, they are of a gross sort." Although this is occasionally true in fisheries, I believe that fisheries biologists have acquired increasing statistical sophistication in recent years. Errors that do occur will be more difficult to detect unless the experimental design is completely and fully specified.

Summary

In recent years, the statistical sophistication of fisheries researchers has substantially increased and, as a result, so has the quality of statistical analysis in published papers. Although the number of elementary statistical errors in published papers such as those found in the *Transactions* are now few, editors and referees need to be on the lookout for more subtle problems associated with sample size and failure to list or test assumptions. The question of assumptions is particularly important in some basic fisheries methodologies, particularly when complicated models or alternative methodology is required.

References

Barlow, J. 1984. Mortality estimation: biased results from unbiased ages. Canadian Journal of Fisheries and Aquatic Sciences 41:1843–1847.

Begon, M. 1983. Abuses of mathematical techniques in ecology: application of Jolley's capture–recapture methods. Oikos 40:155–158.

Burnham, K., D. Anderson, G. C. White, C. Brownie, and K. H. Pollock. 1987. Design and analysis methods for fish survival based on release–recapture. American Fisheries Society Monograph 5.

Chapman, D. G., and D. S. Robson. 1961. The analysis of a catch curve. Biometrics 16:354–368.

Clark, C. W., and M. Mangel. 1979. Aggregation and fishery dynamics: a theoretical study of schooling and the purse seine tuna fisheries. National Marine Fisheries Service Fishery Bulletin 77:317–337.

Cooke, J. G. 1985. On the relationship between catch per unit effort and whale abundance. Report of the International Whaling Commission 35:511–519.

de la Mare, W. K. 1988. On the simultaneous estimation of natural mortality rate and population trend from catch-at-age data. Report to Scientific Committee of the International Whaling Commission, SC/40/01, Cambridge, England.

de Libero, F. 1986. A statistical assessment of the coded wire tag for chinook (*Oncorhynchus tshawytscha*) and coho (*Oncorhynchus kisutch*) studies. Doctoral dissertation, University of Washington, Seattle.

Eberhardt, L. L. 1978. Appraising validity in population studies. Journal of Wildlife Management 42:207–238.

Fisher, R. A. 1951. The design of experiments, 6th edition. Hafner, New York.

Gulland, J. A., editor. 1964. On the measurement of abundance of fish stocks. Rapports et Procès-Verbaux des Réunions, Conseil Permanent International pour l'Exploration de la Mer 155.

Hackney, P. A., and T. E. Linkous. 1978. Striking behavior of the largemouth bass and use of the binomial distribution for its analysis. Transactions of the American Fisheries Society 107:682–688.

Hurlbert, S. H. 1984. Pseudoreplication and the design of ecological field experiments. Ecological Monographs 54:187–211.

Lai, H. L. 1985. Evaluation and validation of age determination for sablefish, pollock, Pacific cod and yellowfin sole; optimum sampling design using age length keys and implications of aging variability in pollock. Doctoral dissertation. University of Washington, Seattle.

Lai, H. L., and D. R. Gunderson. 1987. Effects of aging errors on estimates of growth, mortality and yield per recruit for walleye pollock (*Theragra chalcogramma*). Fisheries Research (Amsterdam) 5:287–302.

MacCall, A. D. 1976. Density dependence of catchability coefficient in the California Pacific sardine *Sardinops sagax caerulea* purse seine fishery. California Cooperative Oceanic Fisheries Investigations Reports 18:136–148.

Paloheimo, J. E., and L. M. Dickie. 1964. Abundance and fishing success. Rapports et Procès-Verbaux des Réunions, Conseil Permanent International pour l'Exploration de la Mer 155:162–163.

Peterman, R. M., and G. J. Steer. 1981. Relation between sport fishing catchability coefficients and salmon abundance. Transactions of the American Fisheries Society 110:585–593.

Robson, D. S., and D. G. Chapman. 1962. Catch curves and mortality rates. Transactions of the American Fisheries Society 90:181–189.

Saville, A., editor. 1980. The assessment and management of pelagic fish stocks. Rapports et Procès-Verbaux des Réunions, Conseil Permanent International pour l'Exploration de la Mer 177.

Seber, G. A. F. 1982. The estimation of animal abundance, 2nd edition. Griffin, London.

Summerfelt, R. C. and G. E. Hall, editors. 1987. Age and growth of fish. Iowa State University Press, Ames.

Ulltang, Ø. 1980. Factors affecting the reaction of pelagic fish stocks to exploitation and requiring a new approach to management. Rapports et Procès-Verbaux des Réunions, Conseil International pour l'Exploration de la Mer 177:489–506.

Writing for Fishery Journals
© Copyright by the American Fisheries Society 1990

Common Statistical Errors in Fishery Research

EDWARD A. TRIPPEL

Department of Zoology, University of Guelph
Guelph, Ontario N1G 2W1, Canada

JOHN J. HUBERT

Department of Mathematics and Statistics
University of Guelph

Abstract.—In a survey of recent papers in fishery journals, we noted four types of common statistical errors: (1) confidence intervals were misused as a test of significant difference between two means; (2) series of *t*-tests were wrongly used to test for differences among several groups; (3) interactions among factors were ignored; and (4) ratios were misused to scale data. These elementary mistakes in statistical practice lead to wrong conclusions about data, and should be avoided.

Fishery scientists must use valid statistical methods to avoid erroneous conclusions from their research. In a survey of papers published in various fishery journals during 1980–1988, however, we observed that authors did not always apply statistical methods correctly. Surprisingly, incorrect use was not restricted to complex statistical procedures that have become accessible in the form of statistical computing software, but rather involved methods discussed in introductory statistics texts. Our objective is to call attention to several of these statistical errors. We hope that their frequency in published works will decline.

We briefly discuss four prominent errors that occur when authors (1) use the overlap of confidence intervals as a test of significant difference between two means, (2) use a series of *t*-tests to test for differences among several groups, (3) ignore interactions among factors, and (4) use ratios to scale data.

Overlap of Confidence Intervals

Whether or not confidence intervals overlap is misused as a statistical method to test for a significant difference between two group means. (Examples are in the *Journal of Fish Biology* 29 [Supplement A]:37, *Transactions of the American Fisheries Society* 112:194, and the *Canadian Journal of Zoology* 65:558.) Because raw data are rarely included in published papers, it is difficult to reanalyze the results; consequently, we use data that have been deliberately constructed to illustrate the issue. A 95% confidence interval (CI) about a sample mean represents the region within which the true population mean falls 95% of the time. Researchers often plot the means and their associated 95% confidence intervals, calculated independently for each treatment level. They then claim that a significant difference exists between each pair of population means whose confidence intervals do not overlap. An explanation of this statistical misapplication (method 1 below) and a suggestion for a valid procedure based on a pooled variance (method 2) follow.

Suppose someone has measured the calcium concentration in the blood sera of rainbow trout from two populations living in lakes of different pH. The issue is to test the null hypothesis, H_0, that the means of the two populations of serum concentrations, μ_1 and μ_2, are equal, that is, H_0: $\mu_1 = \mu_2$. Results yielded the following means ($\overline{Y}$, millimoles per liter) and variances (s^2) for sample sizes (N) of 10 fish in each of the two independent normally distributed samples:

$$\text{population 1 (acid):} \quad \overline{Y}_1 = 1.0, \ s_1^2 = 1.0, \ N_1 = 10;$$

$$\text{population 2 (neutral):} \quad \overline{Y}_2 = 2.2, \ s_2^2 = 1.0, \ N_2 = 10.$$

Method 1.—If only one-sample inferential methodology is used, the hypothesis is rejected when the 95% confidence intervals do not overlap. The lower and upper limits (L_1, U_1) signify the 95% confidence interval for μ_1 and L_2, U_2 is the 95% confidence interval for μ_2. The population means are declared to be significantly different whenever $U_1 < L_2$. For this example, if d is the difference between sample means ($\overline{Y}_2 - \overline{Y}_1$), and $t_c = 2.262$ is the critical value of the t-statistic for 9 degrees of freedom and $P = 0.05/2$ ($P = 0.025$), H_0 is rejected whenever

$$\overline{Y}_1 + t_c(s_1^2/N_1)^{1/2} < \overline{Y} - t_c(s_2^2/N_2)^{1/2};$$

$$\overline{Y}_2 - \overline{Y}_1 > 2.262(1/10)^{1/2} + 2.262(1/10)^{1/2};$$

$$\overline{Y}_2 - \overline{Y}_1 > 0.715 + 0.715;$$

$$d > 1.43.$$

Method 2.—If, instead of calculating a confidence interval independently for each population mean (method 1), we use the pooled variance (s_p^2) to calculate the 95% confidence interval for $\mu_2 - \mu_1$, then $t_c = 2.101$ for 18 degrees of freedom and $P = 0.025$, and H_0 is rejected whenever

$$\overline{Y}_2 - \overline{Y}_1 > t_c[s_p^2(2/N)]^{1/2};$$

$$d > 2.101[1.0(2/10)]^{1/2};$$

$$d > 2.101(0.447);$$

$$d > 0.94.$$

By method 1, the incorrect procedure, the null hypothesis is rejected only when d, the difference between the two sample means of serum calcium concentration in fish sampled from the two lakes, exceeds 1.43. But by method 2, the correct procedure, the null hypothesis is rejected when d is as low as 0.94. Thus, the incorrect procedure yields overly conservative inferences that may lead the researcher to accept the null hypothesis even though it is false. Specifically, whenever d is in the interval 0.94–1.43, the conclusions reached by the two methods disagree. This simple example shows that invalid statistical inferences can be made when tests are based on whether or not two confidence intervals overlap.

TABLE 1.—Probabilities (%) of committing a type I error by use of multiple t-tests in relation to nominal threshold probabilities of type I errors (P_α) and the number of treatments tested two at a time.

Number of treatments	Probability of type I error for P_α =			
	20%	5%	1%	0.1%
2	20	5	1	0.1
3	41	13	3	0.3
4	58	21	5	0.6
10	96	63	23	3.4
20	100	92	52	10.9

Multiple t-tests

Repeated use of the t-test within a single experimental design is often seen in the literature (as in the *Canadian Journal of Fisheries and Aquatic Sciences* 44[Supplement 2]:198). A population ecologist, for example, may wrongly use pairwise t-tests 45 times to test for differences between mean catch-per-unit-effort values for 10 species of a fish community, one test for each possible pair. The t-test is appropriate if conducted just once within an experimental design, but its repeated application leads to the unjustified use of one of the properties of the test—the size of the type I error. The type I error is the probability of falsely declaring a difference between two sample means. This matter of multiple t-tests is a controversial problem, but it has been discussed often; for examples, see Glantz (1980), Duncan and Brandt (1983), O'Brien (1983), and Zar (1984).

If three sample means are tested two at a time with successive t-tests set at P_α = 5%, the probability of incorrectly declaring a significant difference between two extreme means is 13% (Zar 1984); for a t-test with α = 5%, the two-at-a-time approach for equality of 10 means, the likelihood of a type I error is 63%; if there are 20 means, the chance of falsely detecting a difference is 92% (Table 1).

There are alternatives to multiple t-tests. The first alternative is that of analyzing the data by analysis of variance (ANOVA). The only difference between using ANOVA in this way and using t-tests is that the ANOVA pools the error from all groups. This pooling-of-errors method follows from the underlying assumed model for the complete data set. If a significant difference exists, the ANOVA should be followed by a multiple-comparison test to identify which means differ from one another. The standard tests developed by Duncan, Scheffé, or Tukey (Steel and Torrie 1980) serve this purpose.

The second alternative to multiple t-tests is to estimate the realistic probability of a type I error for each two-sample test. To illustrate a conservative approach, multiply the nominal P-value by the number of possible t-tests. In the sample of three means, the tests would be 1 versus 2, 1 versus 3, and 2 versus 3. For a nominal P_α of 0.05 per test, the effective P-value would be 3×0.05 or 0.15, slightly greater than the 0.13 (Table 1) determined by Zar (1984). Thus, if one concludes that there are differences and reports $P < 0.05$, one has about a 15% chance of making at least one incorrect assertion of a treatment effect. So, in this example, one should consider a test to be significant at a nominal P less than 0.05 only if the P-value actually is less than $0.05/3 = 0.016$ (Bonferroni technique; Miller 1981).

TABLE 2.—Mortalities of fish subjected to one of two chemicals (termed I and II in this hypothetical example), in relation to gender.

Gender and chemical	Number tested	Number killed	Mortality (%)
Male			
I	90	18	20
II	10	1	10
Female			
I	20	16	80
II	80	56	70
Genders pooled			
I	110	34	31
II	90	57	63

A third alternative is to carry out each individual test at the level $0.05/[n(n-1)/2]$, when n treatments are compared.

The increased chance of committing a type I error during repetitive testing is not unique to multiple t-tests. A similar situation arises in some environmental studies in which authors compute the Pearson product-moment correlation coefficients among many abiotic and biotic factors (see, for example, *Ecology* 66:1139 and the *Canadian Journal of Fisheries and Aquatic Sciences* 44[Supplement 2]:10). Calculation of pairwise correlation coefficients over many intercorrelated variables eventually yields a significant pair by sheer chance alone. Again, appropriate adjustment should be made before one presents significance levels when this method is used to test hypotheses about abiotic–biotic relationships.

Interactions

Although gender-based differences in morphological and physiological attributes are common among fishes (Scott and Crossman 1973; Bond 1979), researchers do not always consider these differences in data analyses. This results in pooling of such data without statistical justification and may produce misleading results. To illustrate this error, we present a hypothical experiment to determine which of two toxic chemicals causes the greater mortality (Table 2). The data given suggest that, for independently tested genders, the mortality associated with chemical I was greater than it was with chemical II among both males (20% > 10%) and females (80% > 70%). In contrast, chemical II seemed to cause greater mortality than chemical I when gender was ignored (63% > 31%). Thus, one can draw opposite conclusions about the relative toxicities of chemicals I and II simply by deciding to segregate or to pool genders. This statistical phenomenon is referred to as Simpson's paradox (Blyth 1972). Segregation of data is not restricted to gender; it should be considered for any factor (e.g., season) that may affect the variable being measured. Although our example is simple, it clearly exemplifies the danger of oversummarizing data before analysis.

Misuse of Ratios to Scale Data

Researchers frequently try to compensate for the effects of body size by dividing the variable of interest by some measure of size (often body weight), thereby forming such proportion as the gonadosomatic index, the hepatosomatic index, or the rate oxygen consumption per unit weight. (For examples, see *Environmental Biology of Fishes* 5:33 and *American Fisheries Society Symposium* 1:104.) It is difficult to believe that these ratios can be misleading because they are simple to form and seemingly easy to comprehend. A method superior to that of using ratios for removing the confounding influence of varying body size is the analysis of covariance, ANCOVA (Cochran 1957). The technique is based on two procedures: regression and analysis of variance. We illustrate the use of ANCOVA in two simple examples and discuss its superiority over the use of ratios in each. This discussion is adapted from the excellent work on the misuse of ratios in physiological research by Packard and Boardman (1987, 1988).

Increase in Precision

By using ANCOVA, one can increase statistical power and thereby discern differences that might go undetected if the analysis were restricted to ratios. Figure 1A is a bivariate plot showing the distribution of hypothetical fishery data comparing two groups of fish. The variable of interest is gain in body weight and the treatments are two different water temperatures at which fish were reared. The distribution of initial fish body weights was the same in both groups.

Figure 1A suggests that the response of individuals in each of the experimental groups varied linearly with initial body weight. When initial weight is ignored, the values for weight gain can be presented in a dot diagram (Figure 1B). Mean gains were 29.0 g for group 1 and 27.0 g for group 2, and coefficients of variation (100 · standard deviation/mean) are about 20% for each group. When data are examined by analysis of variance, the resultant F-ratio is too small to permit rejection of the null hypothesis; consequently, one would conclude that the means do not differ. In an effort to compensate for variation in body size within each of the samples, the weight gain was divided by the corresponding value for initial weight. When these ratio values are presented in a dot diagram (Figure 1C, in which means are represented by horizontal lines), the precision of the measurements has increased; the coefficients of variation are reduced to about 10%. Despite the increase in precision, however, the F-ratio from an ANOVA still is too small to support an inference of a treatment effect. Additionally, the effects of variation in body weight have not been completely removed from the data, because the ratios still are highly correlated (Pearson product-moment coefficient) with body weight (Figure 1D).

As an alternative to examination of ratios, ANCOVA can be performed on the original data. At the first step in this analysis, straight lines are fitted to the data in each group (Figure 1E). If the slopes of these lines do not differ significantly, an estimate of the "common" slope is calculated (in fact, it is the least-squares estimate). This "common" slope is then used to adjust values for the weight gain; consequently ANCOVA is essentially an ANOVA on these adjusted values. Regressions of adjusted gain in weight versus initial body weight would have zero slopes. The adjusted weight gains for each group are displayed in a dot diagram (Figure 1F). In contrast to Figure 1B, one degree of freedom is lost from the residuals in this analysis as a result of estimating the "common" slope. Variation in the data has been reduced substantially by the adjustment procedure (the coefficients of variation are less than 6%), and the power of the statistical analysis has been increased accordingly. The difference between the two groups now is

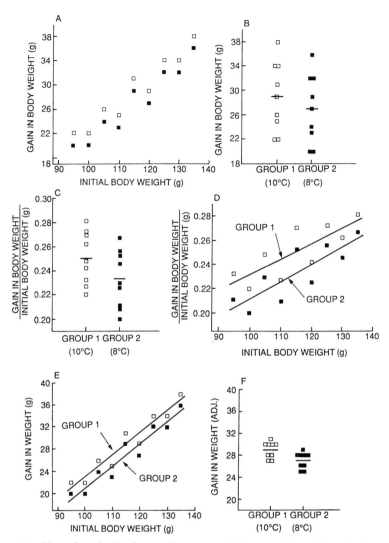

FIGURE 1.—Use of analysis of covariance to gain increased precision in the statistical analysis of weight gain. Fish in group 1 (white squares) were raised at 10°C, fish in group 2 (black squares) at 8°C.

(A) Bivariate plot showing relation between weight gain and initial weight for each group.

(B) Dot diagram showing weight gain, but ignoring initial weight. There is no significant difference between the means (horizontal lines) of the two groups ($F = 0.56$ for 1 and 16 degrees of freedom). The coefficients of variation ($100 \cdot$ standard deviation/mean) are 19.6% for group 1 and 21.0% for group 2.

(C) Dot diagram showing weight gain divided by initial body weight. There is no significant difference between the means of the two groups ($F = 2.70$ for 1 and 16 degrees of freedom). The coefficients of variation are 8.7% for group 1 and 10.1% for group 2.

(D) Bivariate plot showing the relation between the ratio (weight gain divided by initial body weight) and initial body weight. Correlation coefficients (r) are 0.81 for group 1 and 0.84 for group 2; both are significant ($P < 0.01$).

(E) Regression plots of weight gain (G) versus initial body weight (I). For group 1, the estimated regression is $G = -17.0 + 0.4I$; for group 2, it is $G = -19.0 + 0.4I$.

(F) Dot diagram showing the two groups after weight gains have been adjusted for initial body weights by regression. The difference between group means is now significant ($F = 7.5$ for 1 and 15 degrees of freedom; $P = 0.02$). The coefficients of variation are 5.2% for group 1 and 5.6% for group 2.

significant. Therefore, one concludes, after adjustments were made, that the gain in weight differed significantly between the fish reared at the two different water temperatures.

The use of ratios to scale the data for variation in body weight increased the apparent precision of measurements on weight gain, but the statistical analysis of the ratios still lacked sufficient power to discriminate between responses of animals in the two treatment groups. In contrast, the ANCOVA led to an even greater increase in precision of the measurements, and covariance had the added advantage of removing completely the influence of initial body weight on weight gain. The ANCOVA consequently had sufficient power to enable statistical detection of the real effects of the treatments.

If the test for parallelism between regression slopes (Figure 1E) were negative, ANCOVA would not be valid. For the two groups in this example, nonparallel slopes would imply that the treatments had significant effects because they affected the relation between the two variables—weight gain and initial body weight.

Removing Confounding Effects of Body Size

Population ecologists frequently have to test for a significant difference in a response variable between two groups of animals that span different ranges in body size. For example, the distributions of gonad weight in Figure 2A can conceivably occur when two fish populations have different sizes at sexual maturity. The bivariate plot of hypothetical data suggests that gonad weight increases with somatic weight. When the data are displayed in a dot diagram and examined by ANOVA, the average gonad weight of fish in population 2 is significantly greater than that of fish in population 1 (Figure 2B). In an attempt to remove the effect of body size, each gonad weight is divided by the corresponding body weight and the result is expressed as a percentage; the result is the gonadosomatic index. The ANOVA indicates that the mean index is significantly larger for fish in population 2 than it is for fish in population 1 (Figure 2C). Again, one might conclude from this analysis that the groups differ in gonad weight.

As an alternative to using ratios, one can perform an ANCOVA on the original data. Straight lines are fitted to the two sets of data (Figure 2D). When it is determined that slopes of these lines do not differ significantly (in this example they are equal) the "common" slope is used to adjust values for each fish, as was done in the preceding example. The adjusted values are displayed in a dot diagram and examined by ANOVA (Figure 2E). The means for these two populations do not differ once the effects of somatic weight have been removed.

Thus, the use of the gonadosomatic index in this example led to the erroneous conclusion that gonad weights were less for fish in population 1 than for those in population 2. Use of ANCOVA, however, led to the correct conclusion that gonad weights of population 1 were identical to those in population 2 after the effects of size had been removed.

In most studies, no plots of unadjusted data against body size are shown, and authors seldom indicate in the text that bivariate plots were examined. Readers then cannot determine whether the use of ratios led to correct or incorrect conclusions. Formation of ratios is adequate if data vary isometrically. However, the steps that must be followed to justify the use of ratios must also be done for

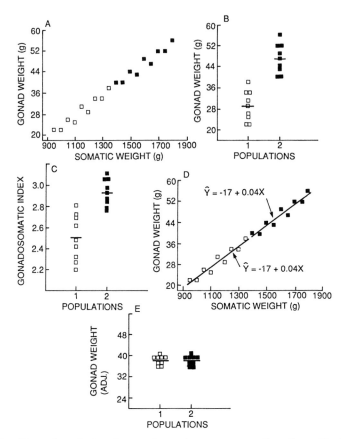

FIGURE 2.—Use of analysis of covariance to remove confounding effects of weight variables.

(A) Bivariate plot showing the similarity of the relation between gonad and somatic weight for populations 1 (white squares) and 2 (black squares).

(B) Dot diagram showing the difference between the two populations when somatic weight is ignored. The difference between the means (horizontal lines) is significant ($F = 45.21$ for 1 and 16 degrees of freedom; $P < 0.01$).

(C) Dot diagram showing the difference between the two populations with respect to the gonadosomatic index ($100 \cdot$ gonad weight/somatic weight). The difference between the means is significant ($F = 25.49$ for 1 and 16 degrees of freedom; $P < 0.001$).

(D) Regressions of gonad weight versus somatic weight for the two populations.

(E) Dot diagram showing the difference between the two populations with respect to the adjusted gonad weights. The difference between the means is insignificant ($F = 0$ for 1 and 15 degrees of freedom; $P = 1$); in fact, the populations are identical.

the ANCOVA. There seems little advantage in scaling data by using ratios, and thus we recommend, along with others (Peters 1983; Donhoffer 1986; Packard and Boardman 1988), that analyses be performed with ANCOVA.

Conclusions

We have identified and discussed four invalid statistical practices common in the fishery literature. These errors can easily be avoided if the following rules are followed.

• Make appropriate statistical plans for data analyses before data are collected. Appropriate planning involves model selection and experimental design. Failure to plan may lead to a substantial waste of time, effort, and money, and it may also inadvertently encourage other investigators to apply the same faulty practices.

• Avoid basing conclusions about differences between two groups on the extent to which confidence intervals overlap.

• Do not use repetitive *t*-tests within the same experiment.

• Do not oversimplify or collapse data by ignoring potentially important interacting factors.

• Do not use ratios to remove the confounding influence of an important factor when ANCOVA is a superior method.

Acknowledgments

This work was supported partly by the Natural Sciences and Engineering Research Council of Canada and by an Ontario Renewable Resources Research Grant from the Ontario Ministry of Natural Resources. We also appreciate the helpful and constructive criticism provided by several reviewers. Victor Kennedy provided the list of selected additional references.

References Cited

Blyth, C. R. 1972. Some probability paradoxes in choice among random alternatives. Journal of the American Statistical Association 67:366–381.

Bond, C. E. 1979. Biology of fishes. Saunders, Philadelphia.

Cochran, W. G. 1957. Analysis of covariance: its nature and uses. Biometrics 13:261–281

Donhoffer, S. 1986. Body size and metabolic rate: exponent and coefficient of the allometric equation. The role of units. Journal of Theoretical Biology 119:125–137.

Duncan, D. B., and L. J. Brandt. 1983. Adaptive *t*-tests for multiple comparisons. Biometrics 39:790–794.

Glantz, S. A., 1980. Biostatistics: how to detect, correct and prevent errors in the medical literature. Circulation 61:1–7.

Miller, R. G. 1981. Simultaneous statistical inferences. Springer-Verlag, New York.

O'Brien, P. C. 1983. The appropriateness of analysis of variance and multiple comparison procedures. Biometrics 39:787–794.

Packard, G. C., and T. J. Boardman. 1987. The misuse of ratios to scale physiological data that vary allometrically with body size. Pages 216–239 *in* M. E. Feder, A. F. Bennett, W. W. Burggren, and R. B. Huey, editors. New directions in ecological physiology. Cambridge University Press, Cambridge, U.K.

Packard, G. C., and T. J. Boardman. 1988. The misuse of ratios, indices, and percentages in ecophysiological research. Physiological Zoology 61:1–9.

Peters, R. H. 1983. The ecological implications of body size. Cambridge University Press, Cambridge, U.K.

Scott, W. B., and E. J. Crossman. 1973. Freshwater fishes of Canada. Fisheries Research Board of Canada Bulletin 184.

Steel, R. G. D., and J. H. Torrie. 1980. Principles and procedures of statistics, 2nd edition. McGraw-Hill, New York.

Zar, J. H. 1984. Biostatistical analysis, 2nd edition. Prentice Hall, Englewood Cliffs, New Jersey.

(Additional references are listed on the next page.)

Selected Additional References

The following papers have helpful discussions of common, simple statistical procedures, experimental designs, or pervasive statistical problems.

Helland, I. S. 1987. On the interpretation and use of R^2 in regression analysis. Biometrics 43:61–69.
Hurlbert, S. H. 1984. Pseudoreplication and the design of ecological field experiments. Ecological Monographs 54:187–211.
Jensen, A. L. 1986. Functional regression and correlation analysis. Canadian Journal of Fisheries and Aquatic Sciences 43:1742–1745.
Kaufmann, K. W. 1981. Fitting and using growth curves. Oecologia 49:293–299.
Laws, E. A., and J. W. Archie. 1981. Appropriate use of regression analysis in marine biology. Marine Biology 65:13–16.
McArdle, B. H. 1988. The structural relationship: regression in biology. Canadian Journal of Zoology 66:2329–2339.
Nielsen, L. A., and W. F. Schoch. 1980. Errors in estimating mean weight and other statistics from mean length. Transactions of the American Fisheries Society 109:319–322.
Peterman, R. M. 1990. Statistical power analysis can improve fisheries research and management. Canadian Journal of Fisheries and Aquatic Sciences 47:2–15.
Pirie, W. R., and W. A. Hubert. 1977. Assumptions in statistical analysis. Transactions of the American Fisheries Society 106:646–648.
Plotnick, R. E. 1989. Application of bootstrap methods to reduced major axis line fitting. Systematic Zoology 38:144–153.
Ricker, W. E. 1973. Linear regressions in fishery research. Journal of the Fisheries Research Board of Canada 30:409–434.
Ricker, W. E. 1984. Computation and uses of central trend lines. Canadian Journal of Zoology 62:1897–1905.
Scheer, B. T. 1986. The significance of differences between means: an empirical study. Comparative Biochemistry and Physiology A, Comparative Physiology 83:405–408.
Schnute, J., and D. Fournier. 1980. A new approach to length-frequency analysis: growth structure. Canadian Journal of Fisheries and Aquatic Sciences 37:1337–1351.
Welsh, A. H., A. T. Peterson, and S. A. Altmann. 1988. The fallacy of averages. American Naturalist 132:277–278.